A Critical Review of Research in

LAND ECONOMICS

by

LEONARD A. SALTER, Jr.

Late Associate Professor of Agricultural Economics
University of Wisconsin

THE UNIVERSITY OF MINNESOTA PRESS
Minneapolis

PRINTED AT THE LUND PRESS, INC., MINNEAPOLIS
MINNESOTA, U.S.A.

LONDON · GEOFFREY CUMBERLEGE · OXFORD UNIVERSITY PRESS

Foreword

Leonard Austin Salter, Jr., was one of the most outstanding and brilliant young men in agricultural economics. His tragic death in the LaSalle Hotel fire in June 1946 was a real loss to agricultural economics and especially to the branch he had chosen as his life's work: land economics.

His death occurred the week before the commencement of the University of Minnesota at which he was to receive his doctor of philosophy degree. This book is the thesis that had been accepted in partial fulfillment of the requirements for that degree, which was awarded *post obitum*.

Leonard Salter, although still a young man, had made a real place for himself in his chosen field. His selection by the University of Wisconsin for appointment as associate professor is an indication of his standing. He had taken a very active and leading part in developing and carrying on research and teaching in land economics, and he drew upon this experience and training in the preparation of this analysis. Expressions received from many workers in the field give strong support to the decision to make it available in published form. Publication was made possible in part by a grant-in-aid from the Farm Foundation of Chicago.

The author examined in detail and reviewed critically a large number of research publications in his study. Any reader who might be disturbed over the frankness of these appraisals will do well to keep in mind the objectives of the author and the basis of his approach as outlined by him in the early part of Chapter IV.

Dr. Salter was denied, of course, the usual opportunity to revise and modify his writing in the process of publication. In the moderate editorial changes which have been made in preparing the material for publication, the constant endeavor has been to preserve the author's ideas inviolate.

O. B. Jesness

Acknowledgments

This dissertation owes much to many friends, teachers, students, assistants, and colleagues I have known and to many authors I have read. But particularly this study reflects the influence and the contribution of the following people, to whom my thanks and obligation are gratefully but inadequately expressed:

Alexander E. Cance, teacher at Massachusetts State College, who introduced me to the social sciences, to the idea of social evolution, and to the place of the social scientist as a contributor to social progress; Irving G. Davis, deceased, teacher and supervisor at Connecticut State College, who had that rare gift of imparting to his students in a few years enough intellectual curiosity to last them a lifetime, and who first opened to me the immense problems of research work in the social sciences; George S. Wehrwein, deceased, teacher and colleague at the University of Wisconsin, who impressed on me some of his boundless interest in those social problems which stem from the allocation of the earth's surface space among men; Oscar B. Jesness, friend and teacher at the University of Minnesota, who gave help and encouragement in many kind ways in bringing my formal training and this dissertation to its present stage of completion; Gertrude C. Salter, my wife, who, at Amherst, Storrs, Madison, and the Twin Cities, has devoted herself to the task not only of making my work possible but of providing an environment that makes even the most arduous work as pleasant as possible; and Dorothy McGuire, who once again has contributed her superb abilities and tremendous energies to putting my work into presentable form.

LEONARD A. SALTER, JR.

May 1946

Table of Contents

CHAPTER I

Introduction

This study, a critical analysis of research in the rural aspects of land economics in the United States, is designed to contribute to the development of scientific method in the social science of land economics.

The process of self-correction, a fundamental attribute of science, tends to become almost automatic in highly developed branches; but in newer fields the need for critical review is more consciously felt. The development of land economics in the United States has been such as to accentuate the need for a review of the research that has been done: Land economics is not only relatively new but has experienced a phenomenal expansion during its brief course.

The tempo of this growth has been set by rapid and sharp social changes during the past few decades. Far from having had a slow and gradual maturation period, land economics research has mushroomed under a demand for results which has been so pressing that only inadequate attention has been given to the reexamination of scientific requirements, to the ordering of knowledge gained, and to the training of scientists. These conditions set the stage for the problem with which this inquiry is concerned.

The existence of the problematic situation is no simple conjecture. There have been marked doubts as to the scientific statement of problems studied and the adequacy of procedures used, in respect to research both in rural land economics and in the broader field of rural social science of which it is a part. Even more striking evidence is known to those who have participated in informal conferences among land economists during the past decade. The proceedings of these conferences are not available for reference, but the workers in the field are not unaware of the confusion which has enshrouded discussions of this subject.

The present attempt to clarify concepts in land economics research and to suggest ways of improving its quality embodies the results of twelve years of continued attention to the methodological problems of land economics, through participation as student, employee, supervisor, and teacher, both in universities and in governmental agencies.

It marks a point of organization in a continuing interest in the progressive evolution of land economics research.

The methods used in making this analysis accord with the principles which this study suggests as essential to the advancement of land economics inquiry. Lack of appreciation of these principles, it is believed, has been responsible for much of the confusion and dissatisfaction which gives relevance to a study like this one. Briefly, the writer has been working on the formulation, testing, and revision of a hypothesis by both operational and conceptual analysis. In this endeavor he has had the opportunity to move from one form of hypothesis to another and from operational to conceptual work on an almost day-to-day basis, and to cover a wide range of existent land economics problems.

It is not now possible to state precisely the author's exact beginning hypothesis. Were such a statement made now, it might be only a conjectural afterthought. But it is not essential that such a phrasing be constructed, for in any problem situation some random probing precedes the formulation of the starting hypothesis. The early guides in the investigation, however, are clear: One is in the classroom and one in the field and laboratory.

One touchstone was provided by a seminar in agricultural economics research conducted under the late Professor Irving G. Davis and his colleagues at Connecticut State College (now the University of Connecticut) in 1933. In this seminar it became evident that there was a very wide range in the quality of published research in rural social science. Also obvious was the great importance of Karl Pearson's *Grammar of Science* as the accepted definitive exposition of scientific method, not only in this seminar, but also among those rural social scientists who did give special attention to general questions of methods of inquiry.

The other touchstone was Professor Davis' investigations of types of farming, a field in which a standardized pattern of inquiry had established itself. As Professor Davis and his staff explored new lines of approach, it became apparent that various means of combining a given set of data gave wholly different results and that the crucial question was not the accuracy of the method in the conventional sense but the concept the researcher had of his problem. This point was repeated and reimpressed during the preparation of the writer's master's thesis.[1] Although this project stemmed directly out of the type of farming work, it was nevertheless necessary to forge a new sys-

[1] L. A. Salter, Jr., *The Place of Part-Time Farming in the Social Economy of a Rural Area,* Master's Thesis (Connecticut State College, 1935).

tem of analysis in order to get at the real substance of the new problem.

In its rough and general form the beginning hypothesis was simply that if research concepts and techniques were altered, then rural land economics research could be more meaningful. The objective in mind was to work toward the development of some general principles of research criticism and methodology; to refine and reformulate the beginning hypothesis so as to construct recommendations specifying the type of modifications needed.

In the decade following the origin of this interest, the writer spent nearly all his time and effort in research, participating in the organization of new research projects, supervising the conduct of field studies in twenty-some states, and correcting and revising the presentation of nearly completed research. As a result of this experience, the limitations of certain types of projects and procedures became more clear and subsidiary hypotheses began to take form. Also, it was possible to test the practicability and effectiveness of some of these ideas in everyday research experience, to discuss their implications with others engaged in similar investigations in different regions, and to spend some time analyzing past research publications.

Later, the writer was able to concentrate on a more thorough study of the literature in land economics research, to discuss these problems with students in fields other than rural social science, to conduct university seminars on research methods, and, at the same time, to carry on further field investigations.

It is under these conditions that the present form of the analysis has been reached. It is out of these experiences as a whole that the questions vaguely raised twelve years ago have taken shape, have been altered in the light of experience, have been tested and observed, and have led to the present critical review of research in rural land economics in the United States.[2]

Some of the original ideas of the writer not only have undergone

[2] See L. A. Salter, Jr., "What Is Part-Time Farming?" *Journal of Farm Economics*, vol. XVIII, 1 (Feb. 1936); "Research and Subsistence Homesteads," *Rural Sociology*, vol. II, 2 (June 1937); "Categories of Land Use Research," *Land Policy Circular* (Feb. 1938); with L. F. Diehl, "Part-Time Farming Research," *Journal of Farm Economics*, vol. XXII, 3 (Aug. 1940); "Land Classification Along the Rural-Urban Fringe," *The Classification of Land* (Missouri Agricultural Experiment Station Bulletin [hereafter, Missouri Agr. Expt. Sta. Bull.] 421, Dec. 1940); "The Problem of Techniques in Land Economics Research," *Proceedings*, New England Research Council (April 1941); "The Content of Land Economics and Research Methods Adapted to Its Needs," *Journal of Farm Economics*, vol. XXIV, 1 (Feb. 1942); "Cross-Sectional and Case-Grouping Procedures in Research Analysis," *Journal of Farm Economics*, vol. XXIV, 4 (Nov. 1942); "A Comment on Deming's Classification of Problems of Inference," *Journal of the American Statistical Association*, vol. XXXVII, 220 (Dec. 1942).

change but have been completely reversed. Others have come into sharper focus so that it now is possible to point to some specific procedural and mental blocks that hinder the improvement of research in rural land economics. Others have so broadened as to suggest that the issues are of fundamental importance not only to land economics inquiry but to social sciences as a whole, and even to science in general. This last implication is not, of course, within the scope of this study; it is hoped that critiques of other branches of inquiry may substantiate, refute, or modify the assertions which appear to be warranted within the present work.

The form of presentation is set, not in terms of chronological experience in making this study, but with a view to weaving together the analysis and the evidence so as to lend perspective and weight to those elements which have been found to be relatively more important for the advancement of research concepts and procedures in land economics.

Each section of this analysis has a function to perform in helping to resolve the basic problem of the study; no part of the material is offered for merely ritualistic purposes. The history of rural land economics in the United States in Chapter II is not just background material but is presented to help define the problems to which research has been applied and the methodological concepts of the researchers who did the work. Chapter III explores the meaning and implications of scientific method. The need for this part of the analysis was not evident early in the study, but it has proved essential to the solution of the problem at hand. The determinations made in Chapter III are important, not only because they provide the basis for generalizing from the studies of specialized problems, but also because they help to explain the direction in which land economics research has developed.

Chapter IV serves as an introduction to the four chapters that follow it and provides in one place certain over-all considerations that should be borne in mind in using research publications as methodological evidence. In each of the four chapters, V, VI, VII, and VIII, segmental fields of land economics are treated. Concentrating upon groups of related reports and upon selected samples in each of these subfields makes possible a more detailed examination of weak and strong points in available studies. In the final chapter an attempt is made to sum up the findings, to tie these up to the generalizations in Chapters II and III, and to present the pattern of suggestions which it is hoped this study supplies to guide future research efforts toward more productive results.

CHAPTER II

The Development of Rural Land Economics in the United States

Land problems have been the subject of discussion and public action since ancient times, but the relevant developments which explain the study of land economics in the United States today date back mainly to the last part of the nineteenth century.

Revolution in Land Policy, 1870–1891

In the history of the United States of America the predominant process under which public land policies have been formulated has been the transferring of an enormous public domain to private ownership and private management.

For just about a century after the American Revolution the basic premises that dominated these land policies were that private action without public interference would assure (1) that the nation's land would be used in such a way as to supply it adequately with raw materials, and (2) that there would be nearly universal family farm ownership. These beliefs, essential parts of the "American Dream," were acknowledged explicitly and implicitly in countless ways; they have been recognized so widely by political and economic historians that they do not need reaffirmation here.

These two tenets were challenged by some federal officials and by a few magazine writers about 1870. In the following two decades they were also called into question by individual scientists and scientific societies. By 1890 the challenge was resounding, and in 1891 Congress gave the first expression of a fundamental change in its national land policies when it passed the famous Revision Act. This act marked the start of a turnabout in legislation and indicated a significant shift away from an attitude of optimism. For the next thirty years land policies were based on an attitude of apprehension — apprehension of a resource famine and of landed monopoly and tenantry. It is in the beginning of this period of revolution in land policies that the roots of rural land economics in the United States are embedded.

The story of this important period in land policy history has been recounted elsewhere,[1] but we are interested in a few generalizations about the economists' position in this development.

So far as the premise of proper utilization of resources is concerned, the first doubts were raised by publicists, nature scientists, and government officials. Among scientific groups the foresters took the leading role, especially after the formation of the American Forestry Congress in 1881 and the American Forestry Association in 1884. Of this agitation Robbins says, "The attack upon the public land system by scientific organizations and prominent periodicals in the early eighties was perhaps unprecedented in the annals of land history."[2] In the same connection, Hibbard remarks that the conservation movement in the United States "owes its origin to the work of science."[3]

The possibilities of maintaining widespread family farm ownership were not doubted so soon; nor were they raised with the same degree of organized pressure as was the case with the assumption of plentiful resources. Agitation and discussion were widespread, however, especially after the census of 1880 reported on the percentage of tenancy and the census of 1890 specifically announced that a frontier line of settlement no longer existed.[4] Certain individuals made much of these facts, notably Henry George, who attained international prominence as a sponsor of land reforms and who, as early as 1879, had called attention to the impending lack of land for free settlement and the growth of farm tenancy.[5]

During this period American economists were, for the most part, concentrating on the deductive theories of the English and the Austrian schools of economics. Beginning in the 1880's, however, a group appeared in this country which had its ties not so much with the English and Austrian as with the German tradition in economics. It was through these German-oriented American economists that the field of economics was attached to the emerging reform in American land policies.

[1] E.g., R. M. Robbins, *Our Landed Heritage* (Princeton, 1942), Chaps. XVII–XIX; B. H. Hibbard, *A History of the Public Land Policies* (New York, 1924), Chap. XXII; C. R. Van Hise, *The Conservation of Natural Resources in the United States* (New York, 1910), pp. 2–14; John Ise, *The United States Forest Policy* (New Haven, 1920), pp. 34–118 *passim.*

[2] Robbins, *Our Landed Heritage,* p. 291.

[3] Hibbard, *Public Land Policies,* p. 472.

[4] J. T. Adams, *The Epic of America* (Boston, 1932), p. 303; F. J. Turner, "The Significance of the Frontier in American History," *Annual Report,* American Historical Association (Washington, 1894), pp. 199–227.

[5] Henry George, *Progress and Poverty* (50th anniversary edition, New York, 1942), pp. 388–94. In 1916, R. T. Ely told the economists that "Henry George is to be praised because he has brought forward the land problem as one of paramount importance." "Landed Property as an Economic Concept and as a Field of Research," *American Economic Review,* vol. VII, 1, supp. (March 1917).

THE ECONOMISTS AND LAND POLICIES

In 1885 the insurgents, led by Professors Richard T. Ely, Simon N. Patten, and E. J. James, established the American Economic Association, largely in protest against the laissez-faire attitude which characterized the English and Austrian schools; and they used the Association to direct economists' attention in this country to pressing questions of public policy.[6]

There were both direct and indirect connections between the insurgent economists and the debate on national land policies. The direct tie was German scientific forestry. It is no coincidence that the leaders in American forestry were German or German-trained and that the upstart economists took advanced work in Germany.[7] In Professor Karl Knies's lectures at Heidelberg, for example, "Forestry was the main feature of conservation developed in the lectures . . . on Practical National Economics and Economics Policy."[8]

The indirect, but certainly no less important, connection was that the German economists stayed close to the concept of political economy at a time when others were tending to center attention on the private, managerial approach. While the German-influenced American economists certainly were not wholly divorced from and free of neo-classical economics, they nevertheless were affected by the German emphasis on (1) the historical or evolutionary aspects of economic behavior, (2) the relationship between law and economics in social institutions, and (3) the importance of current questions of public policy.[9] It is not surprising, therefore, that some of the economists with training in this tradition, at the turn of the century in this country, directed attention to land policy questions. It is of signal importance to stress the fact of this German background, for it made

[6] R. T. Ely, "Conservation and Economic Theory," in R. T. Ely, R. H. Hess, C. K. Leith, and T. N. Carver, *The Foundations of National Prosperity* (New York, 1917), pp. 11–15; G. S. Wehrwein, "Richard T. Ely," *Journal of Land and Public Utility Economics*, vol. XIX, 4 (Nov. 1943); and especially R. T. Ely, *Ground Under Our Feet* (New York, 1938), particularly pp. 132–64 and App. II.

[7] Hibbard says, "The conservation enterprise owes . . . perhaps most of all to Dr. B. E. Fernow" who "had been a forester and conservationist in Germany." *Public Land Policies*, pp. 472–73. Ise says that "to the influence of . . . an official of the Prussian Forestry Department can be traced the meeting of the American Forestry Congress." *U.S. Forest Policy*, p. 95. Robbins states that Carl Schurz "should also be given credit for focusing public attention on the need for a general revision of the whole land system." Robbins, *Our Landed Heritage*, p. 286.

[8] R. T. Ely in *National Prosperity*, p. 19.

[9] See R. T. Ely, *Outlines of Economics* (New York, 1922), App. A; Alexander Gray, *The Development of Economic Doctrine* (New York, 1931), Chap. VIII; John R. Commons, *Institutional Economics* (New York, 1934), p. 115. For historical backgrounds of England and Germany see Karl Polanyi, *The Great Transformation* (New York, 1944), especially Chap. 15.

possible the development of an interest in land economics and affected the approach which land economists took in their work.[10]

In 1890 when the Association for the Advancement of Science petitioned Congress to change the public land system, the American Economic Association showed an interest in the forestry movement. It invited B. E. Fernow to address its annual meeting and issued a special publication on the need of public action in respect to forestry problems, written by Fernow, E. A. Bowers, and Gifford Pinchot—three outstanding leaders of the forestry crusade.[11]

Over the ensuing years the Association promoted discussion of resource and land ownership problems in its publications, and other economic journals also devoted attention to these topics. In addition, during the agricultural depression of the nineties, the economists discussed general questions of farmer movements and farm relief. Even in theoretical debates, land rent became a chief topic of controversy.[12]

In 1892 Ely went from Johns Hopkins to the University of Wisconsin where he proceeded to build a department in which there would be integration of social sciences and a splicing of academic and public interests. As early as 1896, he had Fernow give a course in forestry economics, the first on this continent or in England. Of the early work in land policy problems at Wisconsin, Ely says, "The teaching of conservation at the University of Wisconsin influenced and paved the way for Roosevelt's great conservation movement and the widespread land programs of the present day. . . . I am confident that very soon after I came to Wisconsin in 1892 we began a systematic treatment of what is now called land economics. I treated the whole subject under the awkward title, Landed Property and the Rent of Land."[13]

The Conservation Era, 1891–1921

The thirty years following the Revision Act of 1891 may properly be called the Conservation Era in American history. The fear of a

[10] G. S. Wehrwein, "Institutional Economics and Land Economics Theory," *Journal of Farm Economics*, vol. XXIII, 1 (Feb. 1941), pp. 161–70. Also note discussion of this paper by John Ise, *loc. cit.*, pp. 171–72.

[11] *Publications of the American Economic Association*, vol. VI, 3 (May 1891).

[12] See H. C. Taylor, "Early History of Agricultural Economics," *Journal of Farm Economics*, vol. XXII, 1 (Feb. 1940). Among the specific articles of the time: G. K. Holmes, "Farm Tenancy in the United States," *Quarterly Journal of Economics* (Oct. 1895); A. B. Hart, "The Disposition of Our Public Lands," *Quarterly Journal of Economics* (Jan. 1887); "The Relations between Rent and Interest," *Publications of the American Economic Association*, 3rd ser., vol. V, 1 (1903).

[13] Ely, *Ground under Our Feet*, p. 191. Just what Ely means by having "paved the way" for Theodore Roosevelt's conservation movement is not clear and probably should not be taken too literally. That his work has exerted considerable influence on both early and more recent land policy developments can hardly be doubted.

resource famine and of agricultural tenancy was accentuated by increased immigration, expanded industrialization, and then World War I. The Revision Act, referred to by Hibbard as "the most signal act yet performed by Congress in the direction of a national land policy,"[14] provided, among other things, the original basis for the allocation of some 150 million acres in the public domain as national forest reserves under Presidents Harrison, Cleveland, McKinley, and Roosevelt. Although this program was ended in 1907, four years later Congress provided for additional national forests by actual purchase of private forest lands. The revolution in forest policy was complete.[15]

As early as 1906 some coal lands of the public domain were set aside; but after the National Conservation Congress in 1908, the withdrawal policy was extended to include public lands valuable for oil, potash, copper, phosphates, and other minerals. These actions, all aimed at locking up the nation's mineral storehouse in order to guard it against too rapid and unwise exploitation, were coupled with a series of measures allowing the regulated removal of the minerals under leases. An over-all leasing policy was written into the General Mineral Land Leasing Act of 1920.[16]

In respect to farm land, the Conservation Era was marked by legislative attempts to provide for the homesteading of the less desirable lands that had been avoided during the westward advance of the frontier. In the Desert Land Act of 1877 and in the Carey Act of 1894, Congress made some effort to encourage the cultivation of arid lands under irrigation; but in the Reclamation Act of 1902, a basic change in the approach to irrigation development was made. This act has ever since been the fundamental rock of our national irrigation program.[17] To assist and encourage the establishment of dry farms and grazing homesteads in the arid regions, Congress enacted the Kinkaid Act of 1904, the Enlarged Homestead Act of 1912, and the Stock-raising Homestead Act of 1916.

The second branch of land policy — that having to do with the condition of land ownership — received a good deal of attention in this period, but this interest resulted in only one major stroke of legislation, the Federal Farm Loan Act of 1916. As the percentage

[14] Hibbard, *Public Land Policies*, p. 532. The appraisal is stressed by Professor Paul W. Gates. See particularly his "The Homestead Law in An Incongruous Land System," *American Historical Review*, vol. XLI (July 1936).

[15] *Ibid.*, pp. 530–31.

[16] *Ibid.*, Chap. XXV; Robbins, *Our Landed Heritage*, Chap. XX–XXIII; W. W. Atwood, "The Conservation Movement in America," in A. E. Parkins and J. R. Whitaker, *Our Natural Resources and their Conservation* (New York, 1939), Chap. I; *Proceedings of a Conference of Governors in the White House*, Wash., D.C., May 13–15, 1908 (Washington, 1909).

[17] R. P. Teele, *Economics of Land Reclamation* (Chicago, 1927), p. 69.

of tenanted and mortgaged farms increased from census to census, interest heightened in the possibilities of public aid to help farmers gain more complete ownership of the land. The census of 1910 included a special study of farm mortgages, and in 1913 commissions were sent both by Congress and by private groups to study European programs. After extensive debate the federal land bank system was established by the 1916 legislation.

In this program the "emphasized objectives" were "the checking of the growth of tenancy and the promotion of farm ownership."[18] But rural credit reforms were also tied up with the crusade to conserve resources. It was felt that "the work of replenishing impoverished soils, opening up new fields, and stimulating agriculture in all its branches cannot be long deferred, because the present rate of increase in population is greater than the rate of increase in the means of subsistence, and this youngest among the nations of the earth is in danger of being unable to feed and clothe its people in spite of matchless natural resources. The farmers' debt may be expected to augment at a more rapid progression than in the past . . ." and agriculture will require "enormous funds."[19]

AGRICULTURAL ECONOMICS

During the brisk activity of the Conservation Era, Ely kept in intimate touch with official developments, partly through Charles Van Hise, president of the University of Wisconsin, who was close to Theodore Roosevelt in promoting the nationwide movement. Ely also pushed farther his interest in landed property, both in his teaching and in his writings.[20] Even more important, he encouraged several graduate students to pursue serious work on land policy problems.

One student, Henry C. Taylor, perplexed by the farmer agitation of the 1890's, was anxious to study the economics of the agricultural problems of the day, and in this he was stimulated by Ely and W. A. Scott.[21] Taylor spent a year of academic study in Germany, where

[18] A. G. Black, "Some Current Problems in Agricultural Credit," *Journal of Farm Economics*, vol. XXIII, 1 (Feb. 1941), p. 42; C. W. Thompson, "The Federal Farm Loan Act," *American Economic Review*, vol. VII, 1, supp. (March 1917), p. 124.

[19] M. T. Herrick, *Rural Credits, Land and Cooperative* (New York, 1919), pp. 6–7. (Copyrights dated 1914 and 1915.)

[20] R. T. Ely, *Studies in the Evolution of Industrial Society* (New York, 1903); R. T. Ely, *Property and Contract in their Relation to the Distribution of Wealth* (New York, 1914); Ely *et al., National Prosperity;* A. F. Gustafson, C. H. Guise, W. J. Hamilton, Jr., and H. Ries, *Conservation in the United States* (Ithaca, 1944). Also see papers by Ely and Hess at the Pan American Scientific Congress sessions on "Conservation of Natural Resources" in 1915–1917. *Proceedings of the Second Pan American Scientific Congress*, vol. III (Washington, 1917), pp. 10–21.

[21] H. C. Taylor, "Early History of Agricultural Economics in the United States," Pt. I, unpub. ms. (June 1941) in University of Wisconsin Agricultural Library. See also his "Early History of Agricultural Economics," *Journal of Farm Economics* (Feb. 1940).

he found both rural political economy and farm management taught in the universities. He also traveled and worked in England, where he discovered an existent agricultural economics literature and collected data for a study of English land tenure in which he hoped to find "a clue to the correct understanding of the problems of tenancy and land ownership in the United States."[22] In 1902 Taylor began to teach the economics of agriculture; in 1905 he published a book on agricultural economics, and in the same year he worked in Washington with R. P. Teele on the economics of irrigation.[23] In 1909 Taylor established a department of agricultural economics in the College of Agriculture of the University of Wisconsin.

All this work earned for him the title of "father of agricultural economics" in the United States.[24] To Taylor, a focal point in agricultural economics was the goal of farm ownership. Five chapters—more than half the pages in his textbook—were devoted to land tenure topics, and Taylor's first experiment station bulletin (issued in 1910) dealt with methods of renting farm land in Wisconsin. It was based on information he had collected in 1906. In the next decade came a flurry of farm tenancy studies, which were published in various states.[25]

B. H. Hibbard, who also studied under Ely and spent some time in German universities, later joined Taylor's staff and did a good deal of early work on the growth of farm tenancy in the United States. He continued work in Wisconsin on land tenure problems and embarked on a long-range study of federal land policies, a topic on which Wisconsin students published a whole series of studies. As a member of Taylor's staff and later as Taylor's successor, Hibbard continued, in the Wisconsin department of agricultural economics, active interest in land tenure and land policy research that was, until 1925, allied with work which Ely continued in the department of economics.[26]

[22] H. C. Taylor, *The Decline of Landowning Farmers in England,* University of Wisconsin Bull. 96 (1904), p. 3. Also see his "Early History" (unpub. ms.).

[23] H. C. Taylor, "Economic Problems in Agriculture by Irrigation," *Journal of Political Economy,* vol. XV (April 1907).

[24] Leo Drescher, *Agrarokonomik und Agrarsoziologie* (Jena, 1937), p. 3; S. von Frauendorfer, "Development, Methods, and Results of Agricultural Economics Research in the United States," *Journal of Farm Economics,* vol. X, 3 (July 1928). It is not implied that there is unanimous agreement with this designation. But if agricultural economics and farm management are considered separately, it would be difficult to deny the title to Taylor.

[25] H. C. Taylor, *Agricultural Economics* (New York, 1905). Taylor has written that omission of marketing from this book in both the 1905 and 1919 editions was due to the pressure of other work and the need for haste in publishing. The 1925 edition contained chapters on marketing. See his unpub. ms., especially pp. 92–93; also, H. C. Taylor, *Methods of Renting Farm Lands in Wisconsin* (Wisconsin Agr. Expt. Sta. Bull. 198, 1910). See Chap. VII below for other bulletins which appeared after Taylor's.

[26] See his series of articles on farm tenancy in various regions in the United States

Another who later played a leading role in the history of land economics was Lewis C. Gray. While at Wisconsin he began a historical economic study of southern agriculture and published articles on the economic theory of resource conservation, which have proved to be of lasting influence.[27]

Oliver E. Baker, another of this group, was encouraged to combine his interests in soils and economics and to turn his abilities to the field of agricultural geography and land classification. Baker's work reflected Taylor's interest in the geographical presentation of data on the history of agriculture and Ely's premise that a "sufficient classification" of resources by characteristics and quality was "the first step in any satisfactory treatment of land" beyond the organization of preliminary ideas.[28]

Taylor was responsible for stimulating Charles J. Galpin to study the social aspects of rural community life, and particularly to look into this phase of farm tenancy and ownership questions. As a result of his work in this line, Galpin became the leader of a new branch of rural social science, rural sociology.[29]

In terms of the literature of the profession, all these early trends were capped by the presentation of a paper in 1916 before the American Economic Association by Richard T. Ely, entitled "Landed Property As an Economic Concept and As a Field of Research." In it, Ely argued that although "We are face to face with the gravest economic problems arising out of landed property," available "economic treatises" offer "little to help us in their solution." His discus-

in the *Quarterly Journal of Economics,* vols. 25–27 (1911–1913). Also, see "Farm Tenancy in the United States," *Annals of the American Academy of Political and Social Science,* vol. XL, 129 (March 1912); and Hibbard, *Public Land Policies.* A series of University of Wisconsin bulletins relating to land policies by others than Hibbard appeared around 1910 and included a study of colonial land policy precedents by Ford, railway land grants by Sanborn, educational land grants by Schafer, and the Texas land system by McKitrick. See Chap. V below for Hibbard's contribution in land utilization research work.

[27] L. C. Gray, "Economic Possibilities of Conservation," *Quarterly Journal of Economics,* vol. XXVII, 2 (May 1913), and "Rent Under the Assumption of Exhaustibility," *Quarterly Journal of Economics,* vol. XXVIII, 2 (May 1914); *History of Agriculture in the Southern United States in 1860* (Washington, 1933).

[28] Ely, "Landed Property," *American Economic Review,* vol. VII, 1, supp. (March 1917).

[29] C. J. Galpin, "The Story of My Drift into Rural Sociology," *Rural Sociology,* vol. II, 2 and 3 (June and Sept. 1937); Drescher, *Agrarokonomik, passim.* Also note, C. J. Galpin and E. F. Hoag, *Farm Tenancy: An Analysis of the Occupancy of 500 Farms* (Wisconsin Agr. Expt. Sta. Res. Bull. 44, 1919); and H. C. Taylor, "The Development of Country Life Studies at the U. of Wisconsin," *Rural Sociology,* vol. VI, 3 (Sept. 1941). One outstanding economist who played a role in early agricultural economics developments was Thomas Nixon Carver of Harvard whose work tended to emphasize rural life and welfare issues. Carver was chairman of the committee that set up the Division of Farm Population in the U.S.D.A. in 1920 of which Galpin became the first head.

sion of the concept of landed property emphasized the need for land classification. Then he presented a "rough classification of topics" for research in the field.[30] They were these:

1. Topics connected with terminology.
2. Problems of a historical nature—history of land utilization, history of land policies, history of theories.
3. Other theoretical questions.
4. Conservation topics.
5. Socialization of land—public policy and private and public land ownership and control.
6. Taxation . . . and other public revenues from land.
7. Relation of land to the human element—agricultural labor, farm tenancy, land ownership attainment.
8. Substitutes for free land. New opportunities other than free land.
9. Contemporary land policies, actual and proposed. Enlargement of farm ownership, land reforms, foreign experience.

This paper represented the first formal attempt to round up into an integrated branch of social science some of the areas of work into which economists had moved during the Conservation Era.

THE AGRICULTURISTS

Another development at the beginning of the twentieth century had some influence on the history of land economics research but had, at first, no direct relationship to economics as such.[31]

Under the federal Morrill Act of 1862 and the Hatch Act of 1887, colleges of agriculture and mechanic arts and agricultural experiment stations were established and expanded in the various states, but their work was almost exclusively centered in the physical and biological sciences.[32] Around 1900, however, there were signs of a beginning interest in other aspects of agriculture.

Growing out of the farmers' difficulties in the 1890's, attempts were made at various experiment stations to obtain, by mail questionnaires, farmers' estimates of their costs of producing specific crops. In a few instances, the farmers were asked for their expenditures rather than for a direct estimate of their total costs and profits per acre. At the North Dakota and Minnesota experiment stations W. M. Hays set up cost records for experimental plots. These studies, however, did not yield satisfactory results.

[30] See footnote 28.

[31] Von Frauendorfer, "Agricultural Economics Research," *Journal of Farm Economics*, vol. X, 3 (July 1928).

[32] F. B. Mumford, *The Land-Grant College Movement* (Missouri Agr. Expt. Sta. Bull. 419, 1940), pp. 81–82, 108.

Thomas F. Hunt gave a course at Ohio called "Rural Economy," and his lectures were published in book form. At Iowa State College attention was given to developing a bookkeeping system to determine costs and returns on the college farm. At the University of Minnesota Professors W. M. Hays and Andrew Boss undertook a project in 1902 to determine costs by having farmers keep business records with "route statisticians" who made frequent visits. The first formal results of this cost accounting work were published in 1906.[33]

George F. Warren, a horticulturalist at Cornell University, became one of the outstanding figures in the farm management movement. Having developed a procedure for obtaining orchard information by interviewing farmers rather than by experimentation, Warren extended his work to include the collection of data pertaining to other aspects of farms. From this agricultural survey work in New York,[34] and from the cost account routes developed in Minnesota, grew the field of study known as farm management.

Many members of this group put primary emphasis on the technical agricultural sciences and regarded their task as that of coordinating the subject matter of these fields in terms of the practices of farm operators. They were differentiated from those who felt that agricultural economics dealt primarily with the application of economic principles to agriculture either in terms of the farm or in terms of the agricultural industry. Others regarded farm management as the application of economics to the farm business, and agricultural economics as the application of economics to public issues of agriculture as an industry. It was not until 1919 that the American Farm Management Association, the older organization, and the younger National Association of Agricultural Economists joined to form the American Farm Economic Association—a compromise union of names as well as of academic backgrounds.[35] In various colleges separate farm management and agricultural economics departments had been established, but in later years most of these too joined forces.

[33] Taylor, "Early History of Agricultural Economics," *Journal of Farm Economics* (Feb. 1940); G. F. Warren, "The Origin and Development of Farm Economics in the United States," *Journal of Farm Economics*, vol. XIV, 1 (Jan. 1932); Andrew Boss, "Forty Years of Farm Cost Accounting Records," *Journal of Farm Economics*, vol. XXVII, 1 (Feb. 1945).

[34] G. F. Warren, *An Agricultural Survey* (Cornell Agr. Expt. Sta. Bull. 295, 1911); S. W. Warren, "Forty Years of Farm Management Surveys," *Journal of Farm Economics*, vol. XXVII, 1 (Feb. 1945).

[35] Von Frauendorfer, "Agricultural Economics Research," *Journal of Farm Economics*, vol. X, 3 (July 1928); H. G. Porter, "The Expanding Scope of Agricultural Economics," *Journal of Farm Economics*, vol. XXIII, 1 (Feb. 1941). There is a detailed history of this conflict in Pt. II of Taylor's unpublished manuscript entitled Agricultural Economics from 1908 to 1911 (May 17, 1941). See also Boss, "Farm Cost Accounting," p. 14.

THE U.S. DEPARTMENT OF AGRICULTURE

In 1902 the Bureau of Plant Industry in the U.S. Department of Agriculture established an office to correlate plant science information and to translate it into farm practice recommendations. In the next few years this Office of Farm Management expanded its work beyond farm practices to include studies of farm records, cost accounting, farm equipment, and other farm problems, with the central objective of determining factors that made for financially successful systems of farming.[36]

W. J. Spillman, a mathematician, physicist, and biologist, headed the Office of Farm Management, and under his leadership studies were made of land tenure problems. At first these studies dealt only with tenancy as a system of farming, but later statistical investigations of the progress of farmers toward land ownership (reflecting the nation's concern in its failure to achieve universal family farm ownership) were made.[37]

Spillman further extended the work of his office when he brought O. E. Baker from Wisconsin to begin an atlas of American agriculture and to help formulate land classification categories for census purposes. Baker's work was given added impetus as wartime demands for farm products pushed the margins of cultivation into arid grazing regions and into areas previously cut over for lumber. Spillman reported that "land has become the most important limiting factor in the further development of our agricultural resources. It is therefore time to take stock of our resources in land."[38] In addition to this work in land classification, the Office of Farm Management undertook studies of the problems of managing farms on the fringes of settlement.

In 1917 Spillman presented a paper to the American Economic Association on the work of his office in the field of land classification and tenure.[39] The following list includes all the main items he men-

[36] E. H. Thompson, "The Origin and Development of the Office of Farm Management in the U.S. Department of Agriculture," *Journal of Farm Economics*, vol. XIV, 1 (Jan. 1932); J. T. Horner, "The United States Governmental Activities in the Field of Agricultural Economics Prior to 1913," *Journal of Farm Economics*, vol. X, 4 (Oct. 1928). See also the reports of the federal Secretaries of Agriculture in the U.S.D.A. *Yearbooks* for 1902 to 1909. Also note J. W. Froley and C. B. Smith, *A System of Tenant Farming and Its Results* (U.S.D.A. Farmers' Bull. 437, 1911). For later bulletins following this one, see Chap. VII below.

[37] W. J. Spillman and E. A. Goldenweiser, "Farm Tenantry in the United States," *Yearbook, 1916* (U.S.D.A., Washington, 1917), pp. 321–46. Also see "Papers on Tenancy," *American Economic Review*, supp. (March 1919). In the agricultural economics department at Minnesota, Carl Thompson had started research in farm credit very early, but before his work was published there, he went to Washington where he completed some of the first work in farm credit investigations.

[38] W. J. Spillman, "Work of the Office of Farm Management Relating to Land Classification and Tenure," *American Economic Review*, supp. (March 1918).

[39] *Ibid.*

tioned: (1) The atlas—one section is devoted to land classification and utilization. Another section is devoted to (2) property in land. Under the latter heading the principal subdivisions are land values, land tenures, and land policies. (3) The office has devoted a part of its energies to a study of methods and costs of bringing stump lands into use as farm land. (4) The office made a study of the farm woodlot and of the economics of pasture land. (5) Another problem is the manner in which the farmers acquired their present status of tenant or owner. (6) There are various phases of the tenancy problem, such as the relation of land prices to rental values, percentage of tenancy, length of time required to acquire ownership. The lease contract is considered one of the major problems from the standpoint of farm management. (7) Ranch economics has an important bearing on the settlement of the range country.

This list of topics can be grouped into the following simplified outline:

I. Land resources
  A. Land classification (an inventory of resources)
  B. Land utilization (land management on margins of use)
    1. stump-land farming
    2. arid, semiarid ranching and farming
    3. farm forestry
    4. pastures
II. Landed property
  A. Land ownership
  B. Farm tenancy
  C. Land values

A comparison of this outline of work under way with Ely's outline of suggested topics for research given in the previous year shows that Spillman omitted the academic items, terminology, and history. Also, Spillman's list does not include conservation, which would be entered under "land resources" in the above outline, and taxation, which would be placed under the second category.

THE DIVISION OF LAND ECONOMICS

Obviously, the Office of Farm Management had outgrown itself, at least in name. In 1918 Dr. Taylor was brought to Washington to help reorganize the agency into an Office of Farm Management and Farm Economics. A committee on reorganization suggested that the reorganized agency contain a division to deal with land utilization problems. Its field would include land resources and utilization, land settlement, and land ownership and tenancy. In the next year another committee, which included among others, Taylor, Ely, Baker, and Gray, was established to plan for a division that would take over the land utilization and land tenure work of the former Office of Farm Management. In its report this committee suggested the organization of a division of land economics. This change was made in July 1919, with Dr. Gray in charge of the new division with the new term, land economics, in its title.[40]

The committee's report projected the following five areas of research:

I. Land resources
  A. Classification
  B. Utilization
II. Land values
III. Land ownership and tenancy
IV. Land settlement and colonization
V. Land policies

It is important to note the differences between this outline of work and that presented by Spillman earlier. There are three: (1) the setting out of land settlement and colonization as a separate topic, (2) the separation of land values as a separate topic, and (3) the category of land policies.

The land settlement topic reflects both a change in historical conditions and a difference in research approach. Baker's land classification work had been started during the early years of the war in Europe as increased demands for food supplies and farm land accentuated the fears of a land shortage. Then Spillman's staff also undertook to study the management problems of starting new-land farms. Meanwhile, however, concern over the nation's land and food supply greatly increased, and there was agitation for public programs of land settlement beyond that provided by the Reclamation Act and the modified homestead acts, and for credit aid to farm purchasers beyond that provided by the Farm Loan Act. Several states launched planned colonization projects or rural credit schemes or both. In addition, after this country entered the war, there was further clamor for land settlement aids as a part of a veterans' compensation policy.[41]

[40] *Report of the Committee Appointed by the Secretary of Agriculture* (U.S.D.A. Circular 138, June 1919). Also see Circular 132 (1918); L. C. Gray, *Evolution of the Land Program of the U.S. Department of Agriculture* (U.S.D.A., March 22, 1939); and L. C. Gray, *The Division of Land Economics* (U.S.D.A., Mimeo., Dec. 1925).

[41] See B. H. Hibbard, "The Utilization of Land Not in Farms," and Elwood Mead, "Government Aid and Direction in Land Settlement," *American Economic Review*, vol. VIII, 1, supp. (March 1918); and discussions by R. T. Ely and G. F. Warren in the same issue. Also see reference to Carl Thompson in footnote 37 above.

Under these conditions, there was a need for broader studies that would encompass the whole range of settlement problems. Such questions fit better into the terms of reference of the "political economy" group of agricultural economists than into those of the "farm practice" group of farm management specialists. Ely, Hibbard, John D. Black, and L. C. Gray initiated studies of colonization and credit policies in the northern Lakes States.[42]

Spillman listed questions relating to land values as a subcategory of "property in land," but the committee listed it separately. This difference reflects the changed conditions of the farm real estate market during the war years. In 1916 land values were rising sufficiently fast to be noted as a matter of concern, but their upward spiral in the following years became a matter of exceptional national significance.[43]

The committee's separate category of land policies is not important, for under it was grouped a class of miscellaneous items all of which could have been included under the other four major topics.

Postwar Reorientation, 1921–1931

Following World War I there occurred a basic change in the national land problem, commensurate with that exemplified by the Revision Act of 1891. After a 30-year period of constantly growing fears of food scarcity, an abundance, if not a surplus, came to be the subject of concern. One signal was the sharp reaction in farm prices and the settling back of farm land values. At this time, the Division of Land Economics made an important contribution by taking a long view of the agricultural situation and by calling attention to some deep-seated factors that were at work.

THE 1923 REPORT

Up to 1921 Baker and his staff had been describing the geography of American agriculture and helping to determine land classifications for the 1920 census. Later the division expanded this work to include studies of the forces affecting the nation's needs for land and a determination of the national land requirements. Spurred by the establishment of a departmental Committee on Land Utilization and by the recommendations of President Harding's Agricultural Conference, the division prepared an extensive report on the national land utilization situation. This report, published in the 1923 *Yearbook,* marks a significant turning point in land policy thinking.[44]

[42] See Chap. V below. Also, *Report of the Secretary* (U.S.D.A., Nov. 15, 1919).
[43] See Chap. VII below.
[44] L. C. Gray, O. E. Baker, *et al.*, "The Utilization of Our Lands for Crops, Pasture, and Forests," in *Yearbook, 1923* (U.S.D.A., 1924), pp. 415–506.

The central thesis of the 1923 report was that there was no danger of a shortage of land for agricultural purposes and that a slow and cautious program of farm land development should replace that of eager expansion. Yet it was still felt that there was a threat of a timber shortage and that greater attention should be given to the development of forest resources. This broad view of the nation's land problem, and the facts on which it was based, changed the whole course of land economics in the succeeding years.

The philosophy of this report was embodied in federal legislation in 1924 in respect to irrigation and forestry. Many irrigation projects had run into trouble even before the postwar recession, and their difficulties were accentuated after it. After a report of a special committee known as the Fact Finders, Congress on December 5, 1924, passed an act which required careful advance investigations of the land, the settlers, and the economic feasibility of irrigation projects; and later legislation further revised the participation of the federal government in irrigation undertakings.[45]

Also in 1924, and also through a special committee's report, the Clarke-McNary Act was passed, greatly expanding the legislative basis for direct federal purchase and administration of land for forestry purposes.[46]

During this first half of the 1920 decade, two of five sections in the Division of Land Economics were working on these land utilization problems. One, the land utilization section, had as its "essential objective" the determination of "the proper rate and direction of agricultural expansion and the wise distribution of our land resources between the essential uses crops, pasture, and forests." Another, the section on land reclamation, sale, and settlement, dealt with the more specific problems encountered in pushing "the extension of agriculture to lands now unused"; and it concentrated on irrigation project areas, the cutover regions, and the range areas. In every case, its aim was to help rationalize the progress of agricultural settlement and development on these margins of growth.[47]

These same problems were referred to by Gray in a slightly different way in 1926 when he outlined research needs in the field of land economics in a paper before the American Economic Association. Five

[45] *Extension of the Time of Payment for Settlers on Government Reclamation Projects* (Hearings, House Committee on Irrigation and Reclamation, U.S. Congress, Washington, 1924), especially pp. 4–7. Also, Committee of Special Advisors on Reclamation, *Federal Reclamation by Irrigation* (U.S. Department of the Interior, Washington, 1924).

[46] *Reforestation* (Hearings, House Committee on Agriculture, U.S. Congress, Washington, 1924). Also, M. C. Calkins, "The Clarke Forestry Law," *Journal of Land and Public Utility Economics*, vol. I, 1 (Jan. 1925).

[47] Gray, *The Division of Land Economics*, p. 2.

of the seven topics were (1) land requirements, (2) land classification, (3) forestry economics, (4) grazing, and (5) settlement.[48]

The two remaining topics in Gray's 1926 list, the province of two of the remaining sections of the Division of Land Economics existent in 1925, were land tenure and land values.[49] Between the war and 1925 there were interesting developments in these fields of landed property, although by the end of that time they were overshadowed by the swift pace of events in land utilization. Reference to these topics will be deferred, therefore, until later.

DEVELOPMENTS IN THE STATES

The tremendous about-face in attitudes regarding land and farm production needs after the war had its repercussions all along the line. Reclamation projects felt pinched under their prewar investment commitments, overextended areas of cultivation in the range area were under pressure, the boomer period in the cutover regions was at an end, and in the older highland sections of the East the trend toward farm abandonment became a subject of renewed concern. In these important adjustments, the states involved were not inactive.

In the cutover areas, which had experienced a landward movement until after the war, it became evident that future demands for land would not be sufficient to put most of these lands into farming uses. A start was made in Michigan to inventory the state's land resources in order that prospective settlers would be better able to select the most promising lands first. This program, instigated by the Michigan Academy of Science, came to be called the Michigan Land Economic Inventory. Previous to the war, direct mapping from field observation of land characteristics had been limited to the work of nature scientists. There had long been work in soil and geologic surveying, topographic mapping, forest cruising, and the like. In the social sciences, however, the only related work had been that of Taylor and Baker, whose work had been with generalized secondary data for large areas, and of Galpin, who had done limited work in mapping certain social characteristics of farmers in place.[50]

[48] L. C. Gray, "Land Economics as a Field of Research," *American Economic Review*, vol. XVI, 1, supp. (March 1926).

[49] There was a fifth section in the division, a Farm Labor Section. This unit was created during the war as a separate division when the scarcity of farm labor was a national problem. In 1922 it was reduced to section status and put into the Division of Land Economics. Its importance continued to decline and it soon lost its identity altogether. In the later World War II period the topic revived, but under the Division of Farm Population.

[50] H. C. Taylor, "The Geographical Method," in *The Place of Economics in Agricultural Education and Research* (Wisconsin Agr. Expt. Sta. Res. Bull. 16, 1911); and C. J. Galpin, *Farm Tenancy*, pp. 300–4.

The Michigan land economic survey was "a new departure from the stereotyped piecemeal surveys."[51] It was new primarily because several different nature scientists worked as a team, but it also included an economic inventory of data on land tax and ownership. Lovejoy has stated that the work in Michigan bore out the implications of the 1923 *Yearbook* report and of Dr. Ely's advice to the Harding Agricultural Conference of 1922: "The methods and points of view of the land economists at last dominated and superseded the methods and points of view of the technical agriculturalists who, all these years, had permitted confidence in their technical skill, and undue optimism — if not illusion — as to the economic practicabilities of agriculture, to bolster the assumptions, allegations, and subreptions of professional land-boomers."[52] Yet while the Michigan work advertised the name "land economics" and attracted wide attention to land utilization problems of the cutover areas, still it hardly got beyond field mapping of physical phenomena.

With relative prosperity in the cities during the twenties and increasing agricultural difficulties, cutover lands soon became unattractive on the market and a wave of tax delinquency began to sweep the cutover territories of the Lakes States. This phenomenon, together with the prevailing belief, as stated in the 1923 report, that timber shortages were likely to be acute, added to the interest in the cutover problems. Consequently, in both Wisconsin and Minnesota, studies were made of the tax delinquency situation in the cutover area, special forestry aid laws were passed, and, by the end of the decade, detailed land economic inventories were initiated.[53] The research work in Wisconsin was under the leadership of B. H. Hibbard. On the national level, a broad study of forest taxation was authorized under the sponsorship of the U.S. Forest Service.[54]

The trends against agriculture were felt not only in the areas which had been pushed into farming during the war period and the immediately preceding years; in the older areas of the Northeast the historical

[51] Wade DeVries, "The Michigan Land Economic Survey," *Journal of Farm Economics*, vol. X, 4 (Oct. 1928); R. A. Smith, "The Land Economic Survey in Michigan," *Roosevelt Wild Life Bulletin*, vol. III, 4 (Oct. 1926), p. 683.

[52] P. S. Lovejoy, "Theory and Practice in Land Classification," *Journal of Land and Public Utility Economics*, vol. I, 2 (April 1925).

[53] *Land Economic Inventory of Northern Wisconsin, Bayfield County* (Wisconsin Department of Agriculture and Markets, Bull. 100, 1929); W. A. Duffy, A. R. Whitson, and G. S. Wehrwein, *The Land Economic Inventory of Northern Wisconsin, What It Is and What It Can Be Used For* (Wisconsin Dept. of Agr., Bull. 97, 1929); *Land Economic Survey of Hubbard County, Minnesota* (Minnesota Agr. Expt. Sta. Bull. 317, 1935).

[54] F. R. Fairchild and Associates, *Forest Taxation in the United States* (U.S.D.A. Misc. Pub. 218, 1935).

trend of agricultural recession was accentuated by the same forces that were felt so keenly in the arid and cutover sections. In New York State, for example, farm organization leaders watched the decline in farming in the poorer sections of the state and asked that studies of farm abandonment be undertaken. Following a few spot studies, New York in 1929 passed special legislation for the state purchase and reforestation of rural lands submarginal for agriculture, and two years later authorized an inventory of its rural land resources.[55]

The Division of Land Economics gave some assistance to this work in New York, but it took an even larger responsibility in starting studies of farm abandonment and farm-forest adjustments in areas in Vermont, Pennsylvania, Ohio, Kentucky, and West Virginia.[56]

During these years there was increasing national interest in farm relief, in the existence of heavy surpluses of certain crops, in the depression in land values and the complementary increase in farm foreclosures, and in the relative decline of agricultural prices as compared with industrial prices. By the end of the decade, it was felt in some quarters, including the U.S. Chamber of Commerce, that a reduction in the use of lands submarginal for farming would be instrumental in reducing the severity of the price-depressing effects of crops in surplus. In this way the local land utilization work of the land economists became definitely associated with national farm relief proposals.

In 1931 the Secretary of Agriculture called a national conference on land utilization at which the preceding work and the current proposals were reviewed. The report of this conference[57] stands as a bench mark between the period of reorientation and preparation which followed World War I and the period of feverish public activity which was to begin in the following year.

THE INSTITUTE

No history of land economics during the 1920's would be complete without reference to the Institute of Land and Public Utility Economics. Founded by Richard T. Ely in 1920 at the University of Wisconsin, it was expanded and moved to Northwestern University in

[55] *Report,* New York State Reforestation Commission (Albany, 1932); A. B. Lewis, *Methods Used in an Economic Study of Land Utilization in Tompkins County, New York, and in Other Similar Studies in New York* (Cornell Agr. Expt. Sta., Mem. 160, 1934); F. D. Roosevelt, *Message of the Governor Recommending Survey of the Rural Lands of the State* (New York Legislature, no. 54, 1931). See also Chap. VI below.

[56] L. C. Gray, "Objectives and Methods in the Local Definition of the Extensive Margin in Agriculture," in *Proceedings of the International Conference of Agricultural Economics* (1930).

[57] *Proceedings of the National Conference on Land Utilization,* Chicago, Nov. 19–21, 1931 (Washington, 1932). Also see Chamber of Commerce of the United States, *Marginal Agricultural Land,* Pt. III (1932).

1925, at which time the *Journal of Land and Public Utility Economics* was also launched.

The Institute was a means by which the work of many men and both private and university funds were concentrated on land economics problems. It helped nourish land economics as an academic subject when its chief formal expression was in the name of the U.S. Department of Agriculture's Division of Land Economics. The Institute also expedited the publication of a whole series of books on land economics topics, and it sponsored a number of research investigations. The few rural field studies initiated by the Institute centered on land tenure and were reported in the *Journal*. These studies were conducted at a time when research in land tenure was at a relatively low ebb in the agricultural colleges and in the division.

Probably as important as any aspect of the Institute's influence is the fact that under it Ely held both urban and rural land economics in a common bond. This union is particularly important because by the middle of the 1920's there had been a considerable development in the area of "city planning."[58] In 1926, the same year in which the U.S. Supreme Court upheld city zoning in the famous Euclid-Amber case, the Institute published Harlean James's *Land Planning in the United States for City, State and Nation*. "At first, the book was conceived as a book on city planning" but it developed into a treatise on land planning in general.[59] At the conclusion of the book, Miss James remarked, "The principles of *city planning* which we are now applying to our urban communities are equally applicable to rural regions."[60]

When this statement was made, work on rural land utilization had hardly gone beyond statistical tabulations of specific uses of land and broad estimates of future needs for four categories of rural land—except for the Michigan inventory, which prided itself on being only a coordinated resumé of physical and economic land facts without judgments, evaluations, or recommendations.[61] The land utilization studies which began after the James volume appeared were more specifically aimed to help solve social problems in areas "where the major uses of land are in question."[62] Yet there was still a wide gap between the

[58] Harlean James, *Land Planning in the United States for the City, State and Nation* (New York, 1926), Chap. III.

[59] R. T. Ely, "Preface," in James, *Land Planning*.

[60] James, *Land Planning*, p. 416. In respect to national land policies, the joining of urban and nonurban land economics interests is also to be noted in the report of the Committee on Bases of Sound Land Policy, *What About the Year 2000?* (Harrisburg, 1929), under the sponsorship of the Federated Societies on Planning and Parks.

[61] Lovejoy, "Land Classification," *Journal of Land and Public Utility Economics*, vol. I, 2 (April 1925).

[62] David Weeks, "Scope and Methods of Research in Land Utilization," *Journal of Farm Economics*, vol. XI, 4 (Oct. 1929).

anticipated objectives of these studies and the relatively extreme statement of Miss James. In 1928 Dr. Gray pointed out that many of the land utilization studies "fall short of providing the data necessary for a regional program of land utilization," just as the Michigan inventory lacked a "synthesis of relevant facts involved in forming an economic judgment which may become the basis of action." [63]

In 1929 Professor Weeks commented that the "greatest importance" of these land utilization studies would be "in the local effect upon the social structure and in the value of intelligent regional planning for economic progress." It is worthy of note in this respect that the full name of the Vermont study of land utilization was finally given in 1933 as "Land Utilization as a Basis for Rural Economic Organization." [64] In his summary of land utilization studies of the time, Professor Weeks also noted a "distinctive" project in Marinette County, Wisconsin, in which public officials actually participated in the work. In this study, data were collected "in much less detail" than in other studies, and the information was combined by rough map correlations; but on the basis of this work, the public officials made "recommendations for public action." This Marinette County report came to be the first of a series of reports on land problems in several cutover counties of northern Wisconsin, each report being entitled, "Making the Most of ——— County Land."

PROPERTY IN LAND

Although, as has been noted, land utilization work loomed large in the 1920's, it did not occupy all of the land economists' attention. In fact, when Ely outlined land economics research topics in 1925, he listed land utilization as but one topic, the others being land ownership, land valuation, and land taxation.[65]

The World War I land boom drove prices of land so high that they took more than a decade to settle back to their previous level. This long period of decline was novel in American experience. Following one or two studies of the boom itself, studies of land values and land valuation were carried on all during the 1920's. By the end of the decade, the division had established a standard system for reporting and indexing land prices. The studies of land values after the land boom

[63] L. C. Gray, "The Status of Research Work in Agricultural Land Economics," *Journal of Farm Economics*, vol. X, 2 (April 1928).

[64] Weeks, "Land Utilization," in *Journal of Farm Economics*, vol. XI, 4 (Oct. 1929); C. F. Clayton and L. J. Peet, Vermont Agr. Expt. Sta. Bull. 357 (1933). Also see Chap. VI below.

[65] R T. Ely, "Research in Land and Public Utility Economics," *Journal of Land and Public Utility Economics*, vol. I, 1 (Jan. 1925).

were in line with the general increase in interest in price analysis and mass statistical techniques, which characterized the whole field of agricultural economics during the decade. This work centered primarily in the Division of Land Economics and at the University of Minnesota.[66]

Land tenure research before 1920 grew out of either questions of farm tenant lease contracts or study of the problem of how farm operators achieved ownership of their land. Some of these studies continued for a few years after 1920, but two new lines of interest were also noticeable. The rapid turnover of land during the boom created concern as to the character of the ownership of rented lands; and secondly, land tenure was tied closely into some of the land utilization research, particularly in the farm-ranch areas of the Southwest.

The 1923 *Yearbook,* which carried the landmark report on the nation's land utilization situation, also contained a very comprehensive report on farm ownership and tenancy summarizing all of the findings of the many studies which had already been made.[67] But one must remark that whereas the land utilization report marks the beginning of an intense growth of interest in that subject, the land tenure report is more accurately described as marking the termination of an intensive period of research interest in that subject.

Although Ely always listed land taxation as a special topic of land economics, the Division of Land Economics never embraced it as a distinct line of work for itself. The land utilization work in the Lake States was closely related to problems of taxation; but in the federal department the Forest Service conducted the work in forest taxation. And when late in the twenties the Bureau of Agricultural Economics and various state colleges promoted work in the taxation of farm properties, it was not done by the Division of Land Economics or by men who were regarded as specialists in land economics.

[66] E. H. Wiecking, "Elements of a Cooperative Program for Determining Annual Changes in the Farm Real Estate Situation," *Journal of Farm Economics,* vol. XI, 3 (July 1929). Also see Von Frauendorfer, "Agricultural Economics Research," *Journal of Farm Economics* (July 1928). The start of real interest in price analysis and research is marked by the May 1920 issue of the *Annals of the American Academy of Political and Social Science,* which was devoted to the subject, "Prices" (vol. LXXXIX, 178). On the first page of that issue, Fabian Franklin sounds the keynote: "The enormous advance in prices, which has taken place in the last five years, has caused disturbances, material and psychological, so grave as to be comparable with those directly caused by the waste and destruction of the war itself." In the field of agricultural economics, soon thereafter H. C. Taylor, as chief of the U.S. Bureau of Agricultural Economics, engaged George F. Warren to write a substantial U.S. Department Bulletin on *Prices of Farm Products in the United States* (U.S.D.A. Bull. 999, 1921). Cf. F. A. Pearson and G. E. Brandow, "Agricultural Price Statistics in the United States and Abroad," *Journal of Farm Economics,* vol. XXI, 4 (Nov. 1939).

[67] L. C. Gray, C. L. Stewart, *et al.,* "Farm Ownership and Tenancy," in *Yearbook, 1923* (U.S.D.A., Washington, 1924), pp. 507–600.

RURAL SOCIAL SCIENCE IN THE 1920'S

The decade of the twenties was one of tremendous expansion in the fields of agricultural economics and rural sociology. The pioneer work of men like Taylor, Warren, Hays, Boss, and Spillman had taken hold, and many had become interested in it. The Department of Agriculture established an enlarged Bureau of Agricultural Economics, with whole divisions assigned to farm management, rural sociology, marketing, statistics, land economics, and the like. In addition, the growing severity of farmers' economic problems created widespread debate on all sorts of farm relief proposals. In the midst of all this, the federal Purnell Act made available to the state agricultural experiment stations after July 1, 1925, additional research funds for rural social science work. And soon thereafter the Social Science Research Council began to assign some of its resources to advance the training of professional workers in this field.

Although the outstanding development in agricultural economics in the twenties was the attention given to prices, price statistics, and marketing,[68] nevertheless, farm management continued to receive the bulk of attention. Dr. Taylor reported that in 1926–27 over half the research projects in agricultural economics were in farm management.[69] This solid core of interest in farm management had an influence on land economics in this period, however, and this effect came by way of investigations of type of farming.

If the market mechanism could not be altered to give the farmer better and more stable prices, it might be possible to give farmers greater help in adjusting to forthcoming economic changes. The first requirement of this line of reasoning was that farmers should be provided with advance information on what to expect in the prices of products, services, and resources. This information could be supplied by price economists and statisticians. The second requirement was that farmers should be given general advice as to what kinds of farm organization and management adjustments might best meet the forecast conditions. This was a task for the experts in farm management.

A program intended to fill these needs was launched in 1926 and named the Outlook Program.[70] As a part of it, F. F. Elliott in the

[68] J. I. Falconer, "Survey of Economic Research Now Being Conducted in Experiment Stations," *Journal of Farm Economics*, vol. VIII, 1 (Jan. 1926), pp. 28, 30. Also see Von Frauendorfer, "Agricultural Economics Research," *Journal of Farm Economics*, vol. X, 3 (July 1928); Pearson and Brandow, "Agricultural Price Statistics"; H. C. Taylor, "The New Farm Economics," *Journal of Farm Economics*, vol. XI, 3 (July 1929); and footnote 66 above.

[69] H. C. Taylor, "Research in Agricultural Economics," *Journal of Farm Economics*, vol. X, 1 (Jan. 1928). Cf. Falconer, "Economic Research," *Journal of Farm Economics*, vol. VIII, 1 (Jan. 1926), p. 30.

[70] *Outlook Work: The First 20 Years* (U.S.D.A., Washington, March 1942).

Bureau of Agricultural Economics and economists at several state colleges proceeded to devise means of designating areas within which the systems of farming were sufficiently homogeneous that a determination could be made of the type of economic data needed in the area and of the farm management adjustments that might be applied.[71]

This development was important to land economics for two reasons. First, it meant that farm management research now took on a geographical, mapping approach at a time when land economists were trying to develop methods of land classification and land utilization analysis for the purpose of recommending desirable uses of the land. Secondly, it meant that men who might previously have limited their views to the line fences of particular farms now began to look at the farms of an area as a unit. This approach was capturing the imagination of the land economists at the same time the farm management group first began to feel their way into the problems of typing farms and delineating type-of-farming areas.

Before closing the story of the 1920's, reference should be made once more to the rural sociologists. From its inception, rural sociology was related to the Country Life Movement, which always looked upon rural problems as *community problems*, whether they involved farm tenancy, rural housing, rural schools and churches, or rural landscape design.[72] In many areas, under the Country Life Movement, local farmer committees were established to review the local agricultural situation and to draw up a recommended program of improvements. Sometimes these committees had some or all of their data collection work done by rural sociologists. But in any event, it is important for us to note that rural social surveys in which local laymen participated were carried out even before World War I and were conducted in scattered rural localities from time to time during the decade of 1920.

The New Deal Period, 1932–1942

In the years following 1929 when interest in local land utilization problems was developing rapidly, when some state legislatures were giving attention to remedial measures, and when land utilization research was moving in the direction of community planning, the country was sinking into the depths of the Great Depression. In the winter of 1932–33, with the economy in dire straits, a new federal political ad-

[71] *Ibid.*, p. 10; I. G. Davis, *Types of Farming and Type of Farming Areas in Connecticut* (Storrs Agr. Expt. Sta. Bull. 213, 1936), pp. 122–37; F. F. Elliott, "The 'Representative Firm' Idea Applied to Research and Extension in Agricultural Economics," *Journal of Farm Economics*, vol. X, 4 (Oct. 1928), pp. 495–97.

[72] Country Life issue of the *Annals of the American Academy of Political and Social Science*, vol. XL, 129 (March 1912).

ministration was swept into office with the mandate to harness the remaining power of the country for overcoming its social difficulties.

If this new federal effort, known as the New Deal, can be safely characterized by any one comment, it is that it moved in almost any direction that seemed to offer some hope or promise of possible economic relief — the trial and error method was applied to public policy determination on a vast scale.

In view of the national interest in land utilization at the time (an interest which is conveniently and impressively marked by the proceedings of the 1931 National Land Utilization Conference referred to above), it is easy to see that the new federal policy offered a rare opportunity for the land economists as research students to see some of their recommendations given a trial.

Three of the New Deal's earliest creations were directly related to the interests of the land economists. Under the basic national program of the National Industrial Recovery Act of 1933, funds were assigned through the Public Works Administration to purchase private lands for the enlargement of Indian reservations, federal forests, and wildlife refuges, under the jurisdiction of the respective federal agencies administering such lands. In July of 1934 this program was broadened so that privately owned lands unsuited for agriculture could be purchased for reasons other than the narrow acquisition interests of these agencies. In particular it was intended that the expanded purchase program would help reduce the production of surplus farm products and thus assist in lifting farm prices. Because of the similarity of this objective with that of the Agricultural Adjustment Administration, which undertook to reduce the amount of production on farms that continued as going concerns, the enlarged land purchase program was assigned to a unit known as the Land Policy Section of the Agricultural Adjustment Administration.[73]

A National Planning Board was established in 1933 under the Administration of Public Works to help coordinate the federal public works undertakings. In 1934 this office became the National Resources Board, an executive agency assigned the role of an over-all resources planning and coordinating office for the federal government. This board also encouraged the establishment of similar planning boards on the state and local level. In 1935 the name of the agency was once again changed to the National Resources Committee, but its functions remained the same as previously.[74]

[73] L. C. Gray, "Scope and Objectives of Public Land Acquisition Explained," in *Land Policy Circular* (Nov. 1935), pp. 6–9.

[74] *Progress Report* (National Resources Committee, Washington, June 15, 1936), p. 23.

The National Resources Board established a Land Committee, in part to formulate a policy in regard to the purchasing of land to be removed from farm production on the one hand and the erection of new irrigation projects for bringing land into use on the other. Perhaps the best known work of this committee, however, was the preparation and publication of a series of supplementary reports on various aspects of national land problems and policies. These reports, in a sense, may be regarded as an elaboration and a bringing up to date of the equally famous, but less pretentious, national land utilization report of 1923.

By 1935 the National Resources Board and the Land Policy Section had established a land planning consultant in nearly every state. These men were assigned the particular task of locating and describing areas which might be considered for public purchase out of private (presumably also agricultural) use. In general, they were to help formulate land policies and programs in their respective states.[75]

The National Industrial Recovery Act also contained a specific provision for the use of funds, appropriated under the act, to shift some of the population of areas of limited economic opportunity into other areas where better patterns of living might be developed. A Division of Subsistence Homesteads was set up in the Department of the Interior, and a number of land economists were called upon to help devise plans for the redistribution of overcrowded rural and urban populations. Also, the division, through the Civil Works Administration, launched in 33 states a series of studies dealing with part-time farming. Until this survey was started, only a few scattered states had undertaken research on this phase of rural life, which had been accented by a strong back-to-the-land movement during the early depression years.[76]

Late in 1935 the land purchase and subsistence homesteads activities were regrouped and joined with what had previously been the rural branch of the general federal agencies for family relief to form the Resettlement Administration. In this agency a Land Utilization Division under L. C. Gray was given responsibility for the conduct of the program of submarginal land purchase, and a Land Use Planning Section of this division provided a planning staff for the purchase program and also replaced the National Resources Committee's system of state land planning consultants. The reorganized staff of State Land

[75] *Land Policy Review* (Feb. 1935), pp. 4–5.

[76] "General Information Concerning the Purposes and Policies of the Division of Subsistence Homesteads" (Circular No. 1, Division of Subsistence Homesteads, Washington, Nov. 15, 1933). See also L. A. Salter, Jr., "Research and Subsistence Homesteads," *Rural Sociology*, vol. II, 2 (June 1937).

Use Planning Specialists was charged with conducting general studies and planning activities in the field of land utilization.[77]

In the main, the Land Use Planning Section was a huge expansion of the old Division of Land Economics, geared to include a servicing function for the Resettlement Administration and spread out to influence land program activities in every state. In Washington, the Division of Land Economics, which still existed in name in the Bureau of Agricultural Economics, and the Land Use Planning Section formed an integrated and practically indistinguishable staff.

The Land Use Planning Section included a land values and a land tenure unit as the old division had, but these units were not at first much affected by the new responsibilities of the land economists. The land classification and land settlement units from the old division were very markedly affected by the new developments because they were directly involved in the type of problems to which the Land Utilization Division was devoting its large funds. The Land Use Planning Section also included four other units which had not exisited in the old Division of Land Economics: public finance, legislative analysis, directional measures, and water utilization.

In the work of the Land Use Planning Section, major effort was always given to "planning for adjustments in those 'problem' areas which are characterized by critical maladjustments in the relation of people to land resources." Aside from such work, the section was "primarily directed to serving the needs of the rural resettlement program by locating areas suitable for settlement and closer settlement, and making fiscal analyses of rural resettlement type projects."

The section's "activities in the field of public finance, land occupancy [land settlement], water utilization, institutional arrangements [directional measures], and so forth, are to be looked upon as phases of a larger problem . . . All work which may be undertaken in particular fields should be consistently oriented to the fundamental objective of planning and effectuating constructive adjustments in land use 'problem' areas."[78]

The chief functions of the Land Use Planning Section, then, were to prepare maps showing areas where land was in uses that ought to be discontinued in favor of other uses and to make studies of such areas in order to settle upon a remedial program. These last were known as

[77] *Land Policy Circular* (April 1936), pp. 3–4, 11–18; M. R. Purcell, *A Quarter Century of Land Economics in the Department of Agriculture — 1919–44* (U.S.D.A., Oct. 1945), pp. 12–27.

[78] "Program for Land Use Planning Activities (For the period July 1, 1936 to June 30, 1937)."*Field Instruction LU #32* (Revision 1), Resettlement Administration (Washington, July 10, 1936), p. 1.

area planning studies. The former function was carried out by the state specialists in conjunction with the Washington land classification unit. The latter was performed under the land settlement, or directional measures, unit, sometimes in cooperation with the staff of the public finance unit. The same type of work was done in arid areas, where the provision of water was crucial, under the direction of the water utilization unit.[79]

The public finance unit had the primary task of servicing the operational branches of the overhead agency in their negotiations with public officials in areas where purchase and resettlement work was under way; and the legislative analysis unit concerned itself chiefly with reporting upon and following the progress of state and federal legislative proposals dealing with land problems.

Like the public finance unit, the directional measures unit was not entirely devoted to research but gave a share of its attention to the wider use of public measures, other than purchase and resettlement, for resolving land use maladjustments. In particular it was interested in the development of rural zoning, which had spread across northern Wisconsin and was closely related, in that state, to the land purchase and family relocation work. As previously indicated, land utilization research in Wisconsin had led to the establishment of joint laymen-professional local committees which prepared reports on the land problems of various counties. As the depression deepened, these problems became increasingly acute because of the continued rise of tax delinquencies and the movement of urban unemployed and displaced farmers onto cheap, isolated land in these same counties.

With the store of land use information that was at hand, with the public awareness that had been created by the county committees, and with legislative authority under a permissive 1929 law, in a relatively short time 25 Wisconsin counties enacted zoning ordinances that restricted farm settlement to designated areas. Previous to this time, zoning had been limited in its application to cities and some of their periphery areas.[80] Harlean James's suggestion had become a fact.

Under the Bankhead-Jones Act of July 1937, the land purchase program was given direct statutory support; previously, it had existed by executive orders dealing with the allocation of relief appropriations. With this change, the whole Land Utilization Division was transferred to the Bureau of Agricultural Economics. The next year (as a part of

[79] *Ibid. Passim.*

[80] C. I. Hendrickson, "Rural Zoning: Controlling Land Utilization under the Police Power," and G. S. Wehrwein, "Enactment and Administration of Rural County Zoning Ordinances," *Journal of Farm Economics*, vol. XVIII, 3 (Aug. 1936), pp. 477–92, 508–22.

a larger reorganization which will be discussed later) the operational part of the land program was assigned to the Soil Conservation Service, and the land use planning staff was put back into its old groove in the Bureau of Agricultural Economics as the Division of Land Economics.[81]

COORDINATION THROUGH LAND USE PLANNING

The above-mentioned activities were by no means the whole of the federal agricultural program during the 1930's, nor were they the only ones in which the land economists were involved. They are only those in which the land economists had a dominant role. A brief reference to some of the other rural programs of the federal government is therefore necessary.[82]

One program that made remarkable strides in the expansion of its activities was that which dealt with problems of soil erosion. In 1933 a Soil Erosion Service was established in the Department of the Interior for the purpose of providing field demonstrations of devices and practices of erosion control. In 1935 this agency, renamed the Soil Conservation Service, was given direct legislative status and was transferred to the Department of Agriculture. At the same time, instead of continuing on the slow-moving demonstration basis, the service developed and promoted the establishment of local soil conservation districts under enabling acts passed in the various states. Although economists did not play a large role in this program, except for keeping a check on the farm management results on cooperating farms, there was some growth of interest in the economics of soil conservation.[83]

In 1935 the President was given authority to use public funds to extend electric power lines into unserved rural areas. In 1936 an act of Congress set up the Rural Electrification Administration, which was placed in the Department of Agriculture in 1939. Obviously, decisions as to the location of these projects, especially in marginal areas, involved matters on which the land economists and land use planners were working and on which they were, in some cases, consulted.

During the 1930's several legislative steps were taken in regard to the adjustment of the range areas of the West, in addition to other national agricultural programs which also applied, with modifications,

[81] L. C. Gray, "Our Land Policy Today," *Land Policy Review*, vol. I, 1 (May-June 1938); C. F. Clayton, "The Land Utilization Program Begins Its Second Year," *Land Policy Review*, vol. I, 2 (July-Aug. 1938).

[82] Discussions of all the programs referred to below may be found in "Farmers in a Changing World," in *Yearbook, 1940*.

[83] A. C. Bunce, *The Economics of Soil Conservation* (Ames, 1942), the only full book on the topic published in the United States, is largely a product of his work for the Soil Conservation Service.

of course, to the region. In 1934 the Taylor Grazing Act provided for the control of the remaining public domain lands through the establishment of local grazing districts under the supervision of the Grazing Service of the Interior Department. In 1936 the federal Water Facilities Act provided special funds to help western farmers and ranchers improve their water supply systems.

In 1936 Congress, for the first time, gave an agency other than the U.S. Army Corps of Engineers a hand in planning for flood control. It authorized the Department of Agriculture to report on the extent to which adjustments in land use could assist in providing flood control protection on watersheds in which the Army engineers were conducting their surveys for engineering works. In the department, this assignment was given to the Forest Service, the Soil Conservation Service, and the Bureau of Agricultural Economics, which in turn looked to the Division of Land Economics for leadership in the work.

Reference has already been made to the Agricultural Adjustment Administration. The program of this agency, the central core of the New Deal agricultural activities, was primarily aimed at effecting a rise in the price of agricultural products by restricting their output. After several nationwide attempts to get the professional personnel of the land-grant colleges to help make plans for the reductions and adjustments to be effected, the planning staff of the AAA requested the extension services of the respective states to obtain the answers of farmers in various localities to a list of standardized questions. But even the results of these endeavors did not provide the data the national administrators felt were needed.

Finally an agreement was reached between representatives of the land-grant colleges and the U.S. Department of Agriculture. Under this agreement not only the broad planning of the agricultural programs, but also the coordination of all the many public farm programs, was to be sought through a national system of county and state committees on land use planning, composed of laymen, technicians, and administrators. This program was agreed to in the summer of 1938 after the department had, in 1937, established an Office of Land Use Coordination to help correlate its own activities.[84] By September 1938, the department issued a "County Land Use Planning Work Outline Number 1, Covering an Area Mapping and Classification Project Recommended for County Agricultural Land Use Planning

[84] "Coordination of Department of Agriculture Land Use Planning Activities," *Land Policy Circular* (July 1937); *Circular Letter*, Miscellaneous No. 397, Director of Extension Work, Washington, addressed to all State Extension Directors (July 12, 1938); *Agricultural Planning and Federal-State Relations* (U.S.D.A., Mimeo., undated); Purcell, *A Quarter Century of Land Economics*, pp. 28–33.

Committees." In November the Secretary of Agriculture announced a reorganization of the department with the Bureau of Agricultural Economics as the central planning staff of the department. Within a few months the bureau had in turn established a new Division of State and Local Planning, which became the head office for the program of land use planning and a corps of representatives located in state and regional offices.

The details of the history of the county land use planning effort are not necessary here,[85] except for the fact that the project lasted four years, when, in the midst of war activities, Congress cut off the funds necessary for its continuance.[86]

PROPERTY IN LAND

The depression thirties also witnessed a revival of interest in land tenure, which had received but scant attention since 1923.[87] This development grew largely out of the wave of farm foreclosures that took place in the depths of the depression. Foreclosures on farms had not been uncommon during the twenties as land prices receded from their 1920 peak; but after 1930 they became so prevalent as to create serious social difficulties. Several states passed mortgage moratoria laws; nationally, the Farm Credit Act of 1933 reorganized the federal land bank and associated credit systems and provided for liberal financial resources to help farmers refinance their debts.

Out of these developments came greater attention to questions of farm land ownership, farm real estate indebtedness, and the rise of farm tenancy. Research workers in several states once again took up studies of land tenure and tenancy, particular attention was given by some to land tenure programs in foreign countries, and finally, a President's Committee on Farm Tenancy (under the auspices of the National Resources Committee and with the help of some members of the Land Use Planning — Division of Land Economics — staff) issued a national report on the subject.[88] The recommendations of this com-

[85] See H. R. Tolley, "Contribution of Agricultural Economists to the General Welfare," *Journal of Farm Economics*, vol. XXI, 1 (Feb. 1939); F. F. Elliott, "We, the People," and K. J. Nicholson, "Fore-runners of Unified Programs," *Land Policy Review*, vol. II, 3 (May-June 1939), pp. 1–9, 31–36; M. S. Eisenhower and R. I. Kimmel, "Old and New in Agricultural Organization," and E. A. Foster and H. A. Vogel, "Cooperative Land Use Planning," in *Farmers in a Changing World*, pp. 1125–56.

[86] N. C. Gross, "A Post Mortem on County Planning," *Journal of Farm Economics*, vol. XXV, 3 (Aug. 1943); H. R. Tolley, *The Farmer Citizen at War* (New York, 1943), Chaps. II–V.

[87] Wehrwein refers to the "great flood of land tenancy studies through the second decade of the century . . . Since that time . . . tenancy studies have gone into a period of eclipse." *Research in Agricultural Land Tenure* (Social Science Research Council Bull. 20, 1933), p. 7.

[88] *Farm Tenancy*, Report of the President's Committee, National Resources Com-

mittee led to the passage in 1937 of the Bankhead-Jones Farm Tenant Act, which provided for a liberal credit program to help farm tenants and laborers acquire farms. These events together with the establishment of the Farm Security Administration, which administered the tenant purchase program, further stimulated research interest in land tenure. In addition, the existence of other agricultural action programs, like that of the Soil Conservation Service, dealing with farm problems that were related to tenancy and tenure conditions also gave impetus to research in this field.

With the outbreak of World War II, many of the land tenure research projects, as well as other land economics studies, were laid aside. Lack of personnel and equipment and attention to other more immediate problems of wartime production slowed research work except in the field of land values. Soon after the war began, there was noticeable activity in the farm real estate market and a beginning surge in farm land prices. Mindful of the land price rise of 1914–1918 and of the boom of 1918–1920, with its deflation aftermath, land economists gave considerable attention to developments in the land market and to the possible initiation of land value controls.[89]

Recapitulation

The history of land economics shows it to be a field that has always been close to existent social problems. This characteristic may not be unique, but it is important. In the beginning, it was nature scientists, public officials, and publicists, who gave expression to a growing skepticism regarding the fundamental premises of national policy. They asked whether positive public action were not necessary in order to protect the nation's supply of landed resources and assure widespread ownership of farm land. This great issue attracted the attention of at least one prominent professional social scientist, Richard T. Ely, who recognized its importance and who encouraged many others to give it their attention. The field of land economics has been the result of their work on this problem.

The point of emphasis in the work reviewed has varied with the changing times and with the undulations of public concern. The course of the work has also been affected by developments in the application of social science to other kinds of problems and by minds with training in fields other than social science. Yet the record shows that over

mittee (Feb. 1937). See also H. W. Spiegel, *Land Tenure Policies at Home and Abroad* (Chapel Hill, 1941); and Elizabeth Hooker, *Readjustments of Agricultural Tenure in Ireland* (Chapel Hill, 1938).

[89] M. M. Regan and F. A. Clarenbach, "Land Market Developments and the War," *Journal of Farm Economics,* vol. XXV, 1 (Feb. 1943).

a period of about fifty years, there has continued to be a core of intellectual interest among a group of social scientists who have directed all or a part of their efforts to current phases of the original challenge of the two premises of national land policy. This work has come to be known as land economics.

The intellectual progenitor of land economics was Richard T. Ely. The leader of rural land economics through most of its history to 1939 was Lewis C. Gray. Other men who played a part in setting its direction are Henry C. Taylor and Benjamin H. Hibbard. Gray, Taylor, and Hibbard were students of Ely, and many others who will later be shown to have affected the course of land economics research were in turn trained by Taylor or Hibbard or Ely. Ely's approach to economic problems was not, for his time, traditional. The field of land economics has thus been influenced by those whose training in economics embodied the correlativity of law and neo-classical economics, the historical development of economic society, and the directing force of public action questions on economic inquiry. Other influences came from W. J. Spillman and George F. Warren, both trained in agricultural sciences, and from the emergent work in farm management, rural sociology, and prices and statistics.

The content of land economics can be summarized by reference to three periods: 1900 to World War I, World War I to 1932, and from 1932 to World War II.

1900 TO WORLD WAR I

Problems associated with attaining predominant operator-ownership of farm land, early permeated much of the theoretical and empirical —historical, geographical, and statistical—work of Ely, Taylor, Hibbard, Gray, and Baker. These problems also drew some attention from the sociologists and the farm management specialists. But there was greater interest on the part of farm management men in the managerial side of tenant farming, and they, as well as the land economists, considered the problem associated with contractual relations between landlord and tenants. By the end of World War I, the land boom overshadowed other landed property phenomena, and attention turned to studies of land values and the patterns of land ownership that resulted from this period of rapid transfers.

Problems of land utilization were given considerable attention during this period, but the question of increasing output on existing farms was handled as a farm management and technical agricultural problem. Another category of problems of resource use, however, involved the transfer of nonfarm land into agricultural land. In this study the land economists participated. Ely, Taylor, and Teele early worked on the problems of developing irrigated land when most irrigation work was regarded as an engineering task. Later, Ely and his associates made studies of the agricultural settlement of lands recently cut over for lumber. In addition, Gray did some theoretical work on the general problem of resource conservation, and Baker made a national appraisal of available and potential farm, range, and forest lands.

WORLD WAR I TO 1932

During the 1920's, there were few new developments in research on problems associated with landed property. Land values research begun during the land boom, was affected by the increasing interest in agricultural prices and statistics and was put on a statistical reporting basis. Growing problems of tax delinquency, associated with the shrinkage in rural land values and the cityward drift of the farm population, attracted research attention. This was a contributing factor to the rapid growth of interest in local studies in land utilization. The mounting problem of agricultural surpluses resulted in the outlook program and, in turn, area studies of type of farming, which also contributed to the development of land utilization studies.

Land utilization problems received marked attention in the 1920's. The reversal of the supply-demand situation after the war created a new type of problem — one concerning the fringes of agricultural production. Attention was given to adjustments in decadent, dry-land, irrigated, and cutover areas. At first these studies sought to provide information on the basis of which individuals could make their own decisions more wisely, but by the end of the period, there was a high degree of interest in positive public programs to assist in the conversion of these areas to a less problematic pattern of land use.

1932 TO WORLD WAR II

In the depths of the depression, the fringe areas of agriculture, including the urban-farm belts, were affected by a back-to-the-land movement. This accentuation of difficulties sharpened interest in the problems of these areas. In a sweep of public action, programs were put into effect to ameliorate these problems. Land utilization research became land use planning for these programs, and this activity finally came to be recognized as a possible basis for charting and coordinating a host of general rural-assistance activities. Developments along this line were halted during the war.

The depths of the agricultural depression and the general social reform activities of the period revived interest in questions of farm land ownership and tenancy. This development was slowed by World War II and the accompanying wave of farm prosperity. Another sharp rise in land values was set in motion, however, and predominant research attention was once again turned to problems associated with land values and property transfers.

LAND ECONOMICS IN BRIEF

It is clear from the above that from 1900 to the present, the work that has come to be known as land economics has been directed in the main toward:

1. Problems in the conversion of land from one major use to another general use. (No use to settlement, forestry to settlement, settlement to abandonment, and so on.) These problems, arising out of social changes or social processes, have been looked at by land economists as public rather than private problems.

2. Problems in the attainment of land tenure objectives, whether or not associated with major land use changes. These problems also arise out of social changes, and they have been primarily regarded by land economists as public rather than private problems.

CHAPTER III

Scientific Method and Social Science

Although "no argument should be necessary as to the importance of research method," nevertheless "the whole weight of circumstances is against giving sufficient attention to methodology."[1] This observation, made in 1928 by a committee of eminent agricultural economists, highlights a period when special attention was given to research methodology in the rural social sciences. It may also serve as a theme for the present chapter.

Soon after the funds provided in the Purnell Act began to flow to the states in July 1925, various leaders urged that it was time to examine methodology in the rural social sciences which were to benefit from these grants. At the December 1925 meeting of the American Farm Economic Association, for example, President M. L. Wilson stated that the Purnell Act made "timely the consideration of economic research and the formulation of some statements regarding its objective, relationships, organization, and future trend."[2] And at the same meeting at least ten other main papers were devoted to research appraisal and procedural topics.

The need for such attention to research was owing not only to the anticipated increase in outlays but also to the dissatisfaction felt among members of the profession with past research efforts. Henry Taylor complained that certain types of research procedure were not being used enough;[3] L. C. Gray was critical of empirical research and of available theoretical work;[4] C. L. Holmes felt that farm management research was not yet working along the most fruitful lines;[5] E. W. Allin wondered if all the data collecting that was being done

[1] Advisory Committee on Economics and Social Research in Agriculture, *Research Method and Procedure in Agricultural Economics* (Social Science Research Council, New York, Aug. 1928), p. 2.

[2] M. L. Wilson, "The Source Material of Economic Research and Points of View in Its Organization," *Journal of Farm Economics*, vol. VIII, 1 (Jan. 1926), p. 1.

[3] H. C. Taylor, "Research in Agricultural Economics," *Journal of Farm Economics*, vol. X, 1 (Jan. 1928).

[4] L. C. Gray, "Land Economics as a Field of Research," *American Economic Review*, supp., vol. XVI (March 1926); and "Round-Table Discussions: Land Economics," *American Economic Review*, supp., vol. XVIII, 1 (March 1928).

[5] C. L. Holmes, "Reorganization from the Point of View of the Individual Farm," *Journal of Farm Economics*, vol. VIII, 1 (Jan. 1926).

was properly oriented;[6] and a German observer concluded that, on the whole, American research work did "not present a very convincing picture."[7]

By 1928 the Social Science Research Council had appointed an Advisory Committee on Economic and Social Research in Agriculture; and in August 1928 this committee, composed of J. D. Black, chairman, L. C. Gray, E. G. Nourse, and H. R. Tolley, published a handbook on *Research Method and Procedures in Agricultural Economics*. It was written on the basis of "38 manuscripts submitted by 33 different persons" besides the committee members, excerpts from "perhaps 25 letters," information from "over a hundred correspondents," and reviews and criticisms from "16 different other people."[8] It was, in short, a large-scale attempt to overcome the tendency to minimize attention to methodology, to iron out conflicts in the field, and to stem the criticisms of research administrators who were reported to have "too often seen the old methods produce doubtful or inconsequential conclusions."[9]

Because the 1928 handbook is a composite of the predominant views of the time in respect to research concepts and because it was issued as something like a textbook when rural social science was in a period of considerable expansion, it is important to examine it.[10] Also, the unsettled issues found in the handbook may serve as clues to possible sources of methodological confusion that may arise at other points in this review of rural land economics research.

Even a superficial review of the 1928 handbook indicates that the bulk of attention is given to statistics. Of the 468 pages of the handbook, 287 are devoted to discussion of research methods and 91 per cent of this space deals with the statistical method. In contrast, only a few pages are given to a discussion of broader questions such as the aims, purposes, and general orientation of scientific inquiry, and even these few pages are wholly dependent on Karl Pearson's *Grammar of Science*.[11]

[6] E. W. Allin, "Need for Specific Objectives in Economic Research," *Journal of Farm Economics*, vol. VIII, 1 (Jan. 1926).

[7] Von Frauendorfer, "Agricultural Economic Research," *Journal of Farm Economics*, vol. X, 3 (July 1928).

[8] J. D. Black, "Research Method and Procedure in Agricultural Economics," *Journal of Farm Economics*, vol. X, 4 (Oct. 1928), p. 551.

[9] *Research Method and Procedure*, p. 2.

[10] Schultz says that it was "used extensively." See T. W. Schultz, "Scope and Method in Agricultural Economics Research," *Journal of Political Economy*, vol. 47, 5 (Oct. 1939).

[11] There are 13 references to, and quotations from, *The Grammar of Science* in the first 15 pages of *Research Method and Procedure in Agricultural Economics*. See G. A. Lundberg, *Social Research* (New York, 1929), in which there are five quotations from Karl Pearson in the first ten pages, and note that the subtitle of *Social Research* is "A Study of Methods of Gathering Data." Also see O. C. Stine, *Agricultural Economics—*

The Grammar of Science

An analysis of Karl Pearson's work will serve as a starting point for the subsequent analysis of research methodology for several reasons. (1) As already pointed out, direct quotations from *The Grammar of Science* have been used to present to research workers the fundamental concepts of scientific methods. (2) Even though few research workers may have read Pearson at first hand, *The Grammar of Science* is a well-organized presentation of the dominant scientific beliefs prevailing from 1892 until recently. (3) Pearson is without question one of the great names in statistics, which has so dominated some thinking as to become almost synonymous with research method. (4) Finally, much of Pearson's applied work was in the field of biology, in which many early farm economists had their academic training.

Pearson places emphasis on the classification of facts as the first and foremost step in scientific method; the whole weight of his view of science is on the side of fact-compiling and classification.[12]

The second leg of Pearson's structure is that the scientist finds uniformities of relationships in his data and thus reduces the facts to some brief formula.[13] These uniformities are absolute and invariable in the sense that they form routines of perception which are dependable in the probability of their repetition.[14] When these routines in data are checked by several observers, the resumé becomes a scientific law, the end product of scientific inquiry.

These two points represent the form and the function of science; science consists of the method by which these two tasks of data classification and summarization are carried out. But if one asks how the scientist knows what data to collect and for what purpose he collects and summarizes them, he will fail to find the answers in Pearson's discussion.

At only two points, and in both cases in footnotes, does Pearson recognize the question of what facts to classify. Although Pearson, in support of his own argument, quotes Darwin's report that he (Darwin) "worked on true Baconian principles, and without any theory, collected facts on a wholesale scale," and although Pearson adds that "It is from men like . . . Darwin . . . that we must seek for a true estimate of the Baconian method," nonetheless, in the same section,

A Social Science (U.S.D.A., Mimeo., May 1940), and note its roots in *The Grammar of Science.*

[12] Karl Pearson, *The Grammar of Science* (Everyman's edition, London), pp. 11, 16, 17, 19, 21, 22, 24, 31, 33, 34, 37, 49, 69, 70, 76, 85. "The man who classifies facts of any kind whatever, who sees their mutual relation and describes their sequences, is applying the scientific method . . . " Pearson, *Grammar of Science,* p. 16.

[13] *Ibid.,* pp. 11, 13, 16, 17, 31, 35–37, 73, 76, 77, 83, 85, 87, 94, 97, 98, 101, 118, 120.

[14] *Ibid.,* pp. 112–14, 116–23, 349.

Pearson states in a footnote "That the collection of facts is often largely guided by the imagination as well as the reason must be fully admitted."[15] Later, and again in a footnote, Pearson says that "*classification* is not identical with collection. It denotes the systematic association of kindred facts, the collection, not of all, but of relevant and crucial facts."[16] Beyond these statements, Pearson's concept of scientific inquiry provides no frame of reference for these questions.

Even his references to the aims and purposes of science are not helpful, for his system is a closed one—one that is sufficient within its own method of assembling and summarizing facts. Scientific achievement "does not denote an explanation of the routine of perceptions; it is solely the description of that routine in brief conceptual formulae."[17]

Pearson does point out that beyond the formulation of these "mental shorthand" descriptions of "the sequences of our sense-impressions,"[18] which is science's direct aim, it does have broader claims. It provides good mental training, and it satisfies the esthetic judgment of people—that "insatiable desire in the human breast to resume in some short formula . . . the facts of human experience."[19] In addition, science "can on occasion adduce facts" that bear on social problems, and any of its results may "someday be the starting-point of wide-reaching technical applications."[20]

Although these four claims of science show that Pearson felt that science was a socially useful activity, he provides no functional connection between science and problems, just as in the matter of directing the search for facts, he only vaguely recognizes the insufficiency of his over-all views of scientific method. It may be expected, therefore, that reliance on Pearson for scientific orientation would lead to a predilection for facts and the use of the mathematics of probability as a technique for resuming facts into a brief formula—which fulfills the function of science as Pearson presents it.

The 1928 Handbook

In the 1928 handbook on *Research Method and Procedure in Agricultural Economics,* there are indications both of awareness of the inadequacies of the Pearsonian concept of science and of the confusions that result from a failure to overcome these deficiencies.

[15] *Ibid.*, pp. 33–34.
[16] *Ibid.*, p. 69.
[17] *Ibid.*, pp. 87, 335.
[18] *Ibid.*, p. 98.
[19] *Ibid.*, p. 36. For the first claim, see pp. 13–14.
[20] *Ibid.*, pp. 27, 30.

Although at one point the 1928 committee repeats Pearson's outline in which fact collection and classification are the first step in scientific method, at a later point there is brief reference to what precedes such activity. The committee mentions careful logical analysis of the problem[21] and also refers to the desirability of deductive, qualitative analysis before large quantities of quantitative facts are obtained for classification and summarization.[22]

This preliminary aspect of scientific method receives but little elaboration in the handbook. It involves "*in some cases,* the formulation of an hypothesis . . . that will point the investigation,"[23] "deductive analysis" to get a "clear view of all the factors and forces in the problem" and to think out "possible relationships between them."[24] As a result of this work, "The leaders of the project should have their plan of analytical attack so clearly in mind that they can visualize the sort of tables and graphs that will come out of it."[25]

Beyond this advice, the committee only suggests that research workers ought to review the literature, discuss their projects with general economists, and submit their plans to critics.[26]

It should be noted that only this cursory treatment is accorded these preliminary phases of inquiry even though the committee also states that in "the present stage of development of agricultural economics," these aspects are "more important" than "sound analysis of the data."[27] Yet there is no discussion of what an agricultural economics problem is, or how it is recognized or delineated. Nor is there any discussion of what a hypothesis is, how it works, or what is done with it. Similarly, there is no light thrown on the objectives or purposes of research except in the Pearsonian terms of hunting through data for a brief formula of repetitious facts.

With such an introduction to a discussion of scientific method, the implication clearly is, in spite of a few sentences with different import, that the heart of science is the "collection, summarizing, analysis,

[21] *Research Method and Procedure,* p. 10.
[22] *Ibid.,* pp. 10–14, 28–29.
[23] *Ibid.,* p. 10. Italics inserted.
[24] *Ibid.,* p. 28.
[25] *Ibid.,* p. 30.
[26] *Ibid.,* p. 29.
[27] *Ibid.,* p. 29. In a publication of another committee of the Social Science Research Council dealing with rural sociology, also issued in 1928, it was reported that "In about 72 per cent of the [studies in progress in 1926–1927] some effort was made at an advance formulation or statement of the problem to be studied." While the committee urged greater attention to this task, it is clear that it was more concerned with proper definition of terms and procedures for collecting data than with the character of the problem to be solved. Social Science Research Council, *Rural Sociological Research in the United States* (New York, 1928), pp. 51, 53.

judgment, and presentation of numerical data of mass phenomena."[28] This view is further supported by the treatment of the four methods other than the statistical which together get but nine per cent of the space devoted to all five.[29] In addition, these methods are presented in such a way as to establish a priority scale that runs downward from the statistical.

The experimental method is described as one in which data are secured under conditions in which some "forces" are held "fairly constant" while other "forces" are measured. Basically, this method is only regarded as a tehnique for getting better statistical data. "As contrasted with the statistical method applied to a somewhat larger number of farms in the same area, it would have the advantage of more precision, provided enough farms were included in the experiment. As between a small group handled experimentally and a considerably larger handled statistically, the advantage could easily lie with the statistical attack."[30] The experimental method, then, "for the most part . . . must be looked upon as supplementary to the statistical method, oftentimes an extension or amplication of it."[31]

The informal statistical method is offered as an alternative to the statistical method when the available data are too fragmentary, too few, or otherwise unsatisfactory for formal quantitative handling. It contends for mastery "in the mind of the man in the street." Examples are found in folklore, in the work of philosophies of history, and "the writings of present-day publicists." Yet it is the process by which "generalizations are made from cases." An illustrative example is given in which the method may be classed "either as informal statistics or elementary formal statistics."[32]

The case method is "the study rather much in their entirety of a limited number of units of observation." It is appraised in Marshall's words as "the *intensive* study of all the details" of a few carefully chosen cases. "At its best, it is the best [method] of all; but in ordinary hands it is likely to suggest more untrustworthy general conclusions, than those obtained by the *extensive* method of collecting more rapidly very numerous observations, reducing them as far as possible

[28] *Research Method and Procedure,* p. 36. E. B. Wilson, who, when president of the Social Science Research Council in 1930, found only the methods of definition, mathematics, and statistics as giving merit to the term *economic science.* "Scientific Method in Economic Research," in *Special Lectures on Economics* (U.S.D.A., Mimeo., 1930), p. 6.

[29] This point is also noted by H. R. Tolley, "Recent Developments in Research Method and Procedure in Agricultural Economics," *Journal of Farm Economics,* vol. XII, 2 (April 1930).

[30] *Ibid.,* pp. 314, 319.

[31] *Ibid.,* p. 322. Also see *Journal of Farm Economics,* vol. XI, 4 (Oct. 1928), which contains a series of four articles on the experimental method in economic research.

[32] *Research Method and Procedure,* pp. 309–11.

to statistical form, and obtaining broad averages in which inaccuracies and idiosyncrasies may be trusted to counteract one another to some extent."[33]

The case method has the characteristic, however, of also revealing the "organic unity" of the observed phenomenon, thus giving "a true record of what occurs" in contrast to the statistical method which is "only an abstract approximation."[34]

The lowest order of method is reserved for the method of analogy. Instead of aiming at broad generalization, the analogical method merely tries to search for statements in respect to one instance that can be applied to a second instance. But the handbook was not written with "the intent . . . to inquire closely into the use of analogy" except to warn "against its more obvious hazards."[35]

A review of three research approaches which represent a different form of methodology classification follows these discussions on method. The treatment of these approaches reinforces the comments made above in respect to the hierarchy of methods, the persistent emphasis upon the summarization of mass data into brief formulas as the core of science, and the existence of conflict, doubt, and confusion in respect to questions of critical importance.

In introducing the method of historical analysis, for example, the handbook states that "Ideal historical analysis would . . . surely employ formal statistical method," but "Lacking such data, it *resorts to* informal statistical method; lacking this, to case or monographic studies and even analogies."[36]

Similarly, the section on "qualitative description" is introduced by the remark that "this subject provides for a statistical analysis . . . whenever the necessary data can be obtained"; and it is followed by the statement that "answers to many of the questions posed . . . call for use of data that are at least informally quantitative."[37]

The discussion of qualitative description, written by Dr. E. G. Nourse, emphasizes that this approach looks at economic phenomena in terms of human behavior and activities; it tries to "see how . . . different transactions result in" different events, it gives "a verbal explanation of what is transpiring," it "has given us such understanding as we have of the actual processes by which economic life is

[33] *Ibid.*, p. 300, quoting from Alfred Marshall, *Principles of Economics* (8th edition, New York, 1922), p. 116.

[34] *Ibid.*, pp. 300–1. The terms in quotation marks are from L. L. Bernard, "The Development of Methods in Sociology," *The Monist* (April 1928), p. 314.

[35] *Ibid.*, p. 298.

[36] *Ibid.*, p. 341. Italics inserted.

[37] *Ibid.*, pp. 324, 328.

carrying forward," it fulfills the function "of ascertaining just what is happening and how the several actors . . . are severally performing in the producing of [a] total situation."[38]

That this contribution somewhat baffled the committee is indicated by the subsequent comment that "the handbook, nowhere . . . is as specific as should be as to just how qualitative description is performed. There has been talk enough of the need . . . but little said as to how it is done. . . . No doubt some discussion of the testing of the authenticity of various forms of evidence would be pertinent here. No doubt the principles and procedure of logic are closely involved."[39]

In a similar vein, although the handbook states that ideal historical analysis would be statistical, it also admits that the "method of working with [historical] materials represents a contribution to methodology that has gone further than anything presented elsewhere in the handbook."[40]

Conflicting Social Research Concepts

The confusion that is revealed in the committee's failure to resolve its persistent emphasis on the Pearsonian concept of science and statistical methods on the one hand, and its seemingly contradictory evaluations of preliminary, deductive, qualitative, case, and historical analysis on the other hand, can also be found throughout the literature of social science research.

Professors Odum and Jocher, after accumulating scores of statements on scientific method, nine approaches to social research, six types of social research method, and four types of procedures, observe that "The methods of science simply have not yet been applied to the study of human society"; and they ask for attention to "the need for a new method." Their suggestion is for even greater refinement—a procedure that will reduce social data in "quantitative and particularistic fashion . . . to a *social denominator* or to social constants."[41]

Professor Elmer, like Lundberg, admits that special values may attach to case, historical, and other methods, but tends toward an insistence upon accurately measurable and preferably quantitative

[38] *Ibid.*, pp. 325–27.

[39] *Ibid.*, p. 328. Note the continued discussion of the dilemma in J. D. Black, "Analytical Methods in Agricultural Economics Research," in *Special Lectures on Economics* (U.S.D.A., Mimeo., 1930), pp. 25–36.

[40] *Ibid.*, pp. 341–42. Ten years later, Taylor held that "I am convinced that we have given relatively too much attention to accounting and statistics, and too little attention to the historical method." "The Historical Approach to the Economic Problems of Agriculture," *Journal of Farm Economics*, vol. XIX, 2 (May 1937).

[41] H. W. Odum and K. Jocher, *An Introduction to Social Research* (New York, 1929), Chap. XXIV, p. 305.

correlations. Both review the dispute that has raged between the social case workers with their case histories and the social statisticians. Elmer leans toward the view that the two methods are useful for different objectives, and Lundberg toward the view that case history data, if accurately recorded, provide the basis for quantitative generalization, the *sine qua non* of scientific method. Yet after a vigorous refutation of the critics of statistical procedures, Lundberg states that, "When it comes to practical social work, or research, they rightly demand to be shown the concrete technique whereby case records can be written and analyzed as suggested above. On this point it should be frankly admitted that we are not yet in a position to give a very satisfactory reply. The technique . . . is . . . highly undeveloped."[42]

Some of the stout defenders of nonstatistical procedures argue that because of the nature of social affairs, social science has to settle for these techniques even though they are less scientific. Professor Cooley, for example, puts up a strong recommendation for "Case studies of small institutions" and for "life-study" methods, yet he admits that the methods of statistics are "exact" whereas the case method is a "more fallible procedure of sympathetic observation and interpretation."[43]

Professor Ellwood has argued that the attempt to apply physical science methods in social study is a mistake,[44] while another student of social science method, Dr. Helen Whiteway, goes even further in this direction. After marshalling quotations from Parsons, Mannheim, Weber, Durkheim, Whitehead, Dewey, MacIver, and Sullivan to show the special problems of applying quantitative methods to social subject matter, she concludes that "Exact science and its methods . . . are not equipped to deal with characteristic social problems." Then she goes on to accept Madariaga's statement that in social affairs

[42] Lundberg, *Social Research,* Chap. VIII, p. 183. Cf. M. C. Elmer, *Social Research* (New York, 1939), Chap. VIII, pp. 22–24, and Chap. XXIV. For further case versus statistical arguments, see E. W. Burgess, "Statistics and Case Studies as Methods of Sociological Research," *Sociology and Social Research,* vol. XII (Nov.-Dec. 1927); G. A. Lundberg, "Case Work and the Statistical Method," *Social Forces,* vol. V, 1 (Sept. 1926); C. R. Shaw, "Case Study Method," *Publication of the American Sociological Society,* vol. XXI (1926).

[43] C. H. Cooley, *Sociological Theory and Social Research* (New York, 1930), p. 314 (from "Case Study of Small Institutions as a Method of Research," *American Sociological Society Publications,* vol. XXII [1928], pp. 123–32); *ibid.,* p. 331 (from "The Life-Study Method as Applied to Rural Social Research," *Publication of the American Sociological Society,* vol. XXIII [1929], pp. 248–54); and *Research Method and Procedure,* p. 301 (from C. H. Cooley, "The Roots of Social Knowledge," *The American Journal of Sociology* [July 1926], p. 73).

[44] C. A. Ellwood, "Scientific Method in Sociology," *Social Forces,* vol. X, 1 (Oct. 1931), p. 15; see also R. M. MacIver, "Is Sociology a Natural Science?" *Publications of the American Sociological Society,* vol. XXV, 2 (May 1931).

one will find "divination more useful than observation and feeling than logic."[45]

Professor Sorokin puts the discussion on a somewhat different basis in suggesting that the quantitative procedures of "causal-functional" study are useful for lower orders of inquiry, whereas in the search for greater generalizations concerning order and unity in the higher realms of cultural values, only a "logico-meaningful" method of analysis is applicable.[46]

Professor Landis feels that in the study of human behavior, as contrasted with the natural sciences, special methods of analysis are necessary; and that there are only limited possibilities of developing broad generalizations because of the critical importance of human interaction.[47] In the field of psychology the same theme of interaction among the elements of a situation has led Professor Kohler to rebel against common statistical procedures; and more recently Professor Thurstone has developed a "factor analysis" technique to try to preserve interactions in his studies of human traits.[48]

In general economics, Professor Commons also asked for "a new logic, a new methodology, even a new terminology" that would be better able to deal with social behavior than the currently quantified procedures.[49]

A recent demand for a re-evaluation of the predominant concepts of scientific method in social science and its emphasis on mass quantitative data has been made by Professor Lynd. His position is relevant to this study, not just because like so many others (including the 1928 handbook editors) he sees the fallacy of the unqualified "Baconian myth," which establishes fact-gathering as the crux of science.[50] He is not alone, as has been shown, in emphasizing that

[45] Helen L. Whiteway, *Scientific Method and the Conditions of Social Intelligence* (St. John's, 1943), pp. 38–39, 47 (from Salvador de Madariaga, *Englishmen, Frenchmen, Spaniards* [Oxford, 1931], pp. xv–xvi). Miss Whiteway also maintains that social inquiry must be "largely subjective" and that it deals with elements that have to be "defined by instinct and felt by intuition." Pp. 27–28.

[46] P. A. Sorokin, "Forms and Problems of Culture Integration and Methods of Their Study," *Rural Sociology*, vol. I, 2 and 3 (June and Sept. 1936).

[47] P. H. Landis, "The Concept, Social Process: Its Meaning and Usefulness in the Study of Rural Society" (Mimeo., 1940), p. 1. An abstract of this paper is in *Rural Sociology*, vol. VI, 1 (March 1941).

[48] Wolfgang Kohler, *Gestalt Psychology* (New York, 1929), *passim;* and L. L. Thurstone, "Factor Analysis as a Scientific Method with Special Reference to the Analysis of Human Traits," *Eleven Twenty-Six* (Chicago, 1940), pp. 78–105.

[49] J. R. Commons, "Evaluating Institutions as a Factor in Economic Change," in *Special Lectures on Economics* (U.S.D.A., Mimeo., 1930), pp. 7–22.

[50] Morris R. Cohen, "Method, Scientific," in *Encyclopedia of the Social Sciences,* vol. 10 (New York, 1933), pp. 390–91. Also see M. R. Cohen and E. Nagel, *An Introduction to Logic and Scientific Method* (New York, 1934), p. 215; W. E. Spahr and R. J. Swenson, *Methods and Status of Scientific Research* (New York, 1930), pp. 263–64; A. A. Young, "Economics," in Wilson Gee, ed., *Research in the Social Sciences* (New York, 1925), p. 67.

social research must deal with human behavior as such while mass statistical methods deal with people "at several removes with their differences [erroneously] assumed to 'cancel each other out.'"[51]

The special importance of Lynd's view is, as may be gathered from the very title of his book, *Knowledge for What?*, that he emphasizes the question of purposes in social inquiry. Except insofar as the objectives of inquiry are stated to be the determination of factual generalizations, laws, principles, or numerical relationships, this fundamental and seemingly obvious issue is slighted in many discussions of social science methods.[52] Without attention to the objectives and purposes of inquiry, how is it possible to appraise these issues of method and procedure in research?

In his recent work, Lynd concerns himself with the gap between social science and existing social problems: how to make social sciences more useful "instruments for the analysis of [society's] more critical problems and for the devising of indicated concrete programs of action." It is in response to this query that he comes to emphasize the importance of working out research procedures that properly treat the "behavior of people that provides the dynamics of change."[53]

Professor Landis also sees that the "social process" approach he advocates takes on importance in social science when science is seen as a "problem-solving device." Yet he carefully states that "solving" is used in the "sense of understanding, not in the sense of ameliorating undesirable social conditions."[54] Although Landis does not go on to indicate what social process research requires in the way of procedures of inquiry, Professor Cooley does argue a form of the case method—even though it is more fallible than the statistical because it has a greater potentiality "to extend knowledge of what is going on . . . and . . . *power to control the process.*"[55]

Professor Lindeman also insists on the requirement that social research should "somehow be usable as an implement of social change." He sees that historical, logical, analogical, and statistical methods have varying utilities. He proposes the addition of a psychological approach, the use of participant-observer techniques, and the synthesis of these procedures. To accomplish the synthesis, he suggests that "if six in-

[51] R. S. Lynd, *Knowledge for What?* (Princeton, 1940), pp. 25–26, 30–31. Also see Commons, "Evaluating Institutions," in *Special Lectures on Economics* (U.S.D.A., Mimeo., 1930).

[52] See Lundberg, *Social Research;* Pearson, *Grammar of Science;* Elmer, *Social Research;* Odum and Jocher, *Introduction to Social Research; Research Methods and Procedure;* Spahr and Swenson, *Scientific Research;* and other works previously cited.

[53] Lynd, *Knowledge for What?*, pp. 9, 46. See also pp. 25–26 and 37–40.

[54] Landis, "Social Process," p. 3.

[55] Cooley, *Sociological Theory,* p. 322. Italics inserted.

vestigators, all attempting to study a single situation in terms of an accepted set of analytical categories, were to function cooperatively, each separate skill would supplement the other and the net result would be a set of facts and interpretations immediately usable for purposes of social change."[56]

Lynd deals more with the orientation of the various social sciences than with specific research methods, but he does insist that the empirical method of the natural sciences "has imported into the social sciences an undue emphasis upon certain kinds of assumed hidden orderliness and impersonal causality in the datum of social science, which it becomes the task of the scientist simply to discover and to describe."[57]

That such importation has been made in the rural social sciences and in social sciences generally is abundantly clear in the 1928 handbook report, in Pearson's *Grammar of Science,* and in the purposes of science as given by many of the authors referred to above. In economics, Allyn Young argued that science's "first creed" is dependable regularity and uniformity and that economic science sees society as "constituting an intricate but reliable mechanism operating in an orderly and predictable way."[58]

In contrast, Professor Frank Knight points out that since people are the subject matter of social research, this goal of describing uniformities for purposes of controlling change is not directly transferable from natural to social science.[59] The objective of social science, says Lynd, is not just to find such uniformities as may exist in society, but "If such order is to exist in culture, it must be *built into it by science,* and not merely discovered in it."[60]

The idea that social science should be aimed at problem-solving rather than at the discovery of brief resumés of uniformities in data is also supported by Professor John Dewey. But he also points out that any sharp distinction between social and natural science in this regard involves a misconception of the task and methods of science in

[56] J. J. Hader and E. C. Lindeman, *Dynamic Social Research* (New York, 1933), pp. ix, 266, *et passim.* Also, E. C. Lindeman, *Social Discovery* (New York, 1924), *passim.*

[57] Lynd, *Knowledge for What?,* pp. 123–24.

[58] Young, "Economics," in Gee, ed., *Research in the Social Sciences,* p. 55. Also note: "Without such uniformity of nature, it may be said, everything would be chaotic; there could be no knowledge." *Research Method and Procedure,* p. 276. Likewise, Pearson, "Anything . . . that tends to weaken our confidence in the uniform order of phenomena . . . tends also to stultify our reasoning faculty by destroying the sole basis of knowledge . . . Conviction, therefore, of the uniform order of phenomena is essential to social welfare." *Grammar of Science,* p. 120.

[59] F. H. Knight, "Fact and Interpretation in Economics," in *Special Lecture on Economics* (U.S.D.A., Mimeo., 1930), pp. 37–45.

[60] Lynd, *Knowledge for What?,* p. 125. See also L. K. Frank, "The Principle of Disorder in Economic Affairs," *Political Science Quarterly* (Dec. 1932), quoted in Lynd, *loc. cit.*

practice. In the same lecture series in which Young was putting the search for order as the aim of both natural science and economics, Dewey pointed out that such an objective had long been outmoded even in the natural sciences, although "To have denied, *at one period,* the existence of such uniform stages of succession would have seemed to be the denial of the possibility of any social science whatever."[61]

Because this position of Dewey's implies a reoriented concept of science in general, it should receive attention in this paper as a possible mode of resolving the conflicts that are apparent. Before entering upon it, however, it will be well to make the point that since the 1928 report on research methods and procedures in agricultural economics, a major source of dispute has been this same question which has been shown to exist in the general social science literature—namely, the relevance of social research to social problems.

Research and Action

After the advent of the New Deal and its enormous expansion of public action programs, there was a renewed wave of interest in the role, the methods, and the achievements of rural social science research. This interest centered around the criticism that research was not geared up to the solution of existing social problems. This feeling was most forcibly expressed as a major consideration in the reorganization of the Bureau of Agricultural Economics in 1938–39. Howard Tolley, then chief of that bureau, specifically stated that the reorganized bureau was charged with three functions: (1) "fundamental research," a "good portion" of which will be directed "more specifically toward immediate planning and operating problems"; (2) cooperation in "The development of plans and suggestions for agricultural programs"; and (3) cooperation in "developing . . . an integrated and continuing national agricultural program."[62]

In the field of land utilization research, Ernest Wiecking urged that "more of our research will now need very definitely to be oriented to the action point of view."[63] Professor Murray R. Benedict also saw

[61] John Dewey, "Philosophy," in Wilson Gee, ed., *Research in the Social Sciences,* p. 256. See also B. Ginzburg, "Science," in *Encyclopedia of the Social Sciences,* vol. 13 (New York, 1934), p. 602; Commons, "Evaluating Institutions," in *Special Lectures on Economics* (U.S.D.A., Mimeo., 1930); and Felix Kaufmann, *Methodology of the Social Sciences* (New York, 1944), Chap. XIII.

[62] H. R. Tolley, "Contribution of Agricultural Economics to the General Welfare," *Journal of Farm Economics,* vol. XXI, 1 (Feb. 1939), pp. 16–17. See also Mordecai Ezekiel, "The Broadening Field of Agricultural Economics," *Journal of Farm Economics,* vol. XIX, 1 (Feb. 1937).

[63] E. H. Wiecking, "Application of Research to Action Programs," *Journal of Farm Economics,* vol. XIX, 2 (May 1937), p. 596. See also M. L. Wilson, "The Problem of Poverty in Agriculture," *Journal of Farm Economics,* vol. XXII, 1 (Feb. 1940), pp. 22–23.

"demands for new types" of land use studies, because of the new problems of "finding out what we want to do" and the lack of "a well-developed methodology adequate to present day needs."[64]

In a similar vein, land tenure research was found wanting insofar as public action was concerned. In the words of James Maddox, "If research workers can be roused from their lethargy of objectivity to the point of critically examining their own methods in view of the demands made by policy-making, it may yet be shown that systematic observing, objective analysis, and reflective thinking can significantly supplement" or "take the place of . . . 'common sense'"; but so far research was found to have contributed next to nothing in respect to important public policy questions in land tenure.[65] Professor Wiley also reported that "the biggest tenure *problem* in the South today is to recognize one when we see it, and the biggest *research need* is to know how and what research to do."[66]

These criticisms were not limited to land economics as a branch of rural social science. In the field of farm management, "which claims more personnel and resources than any other branch of agricultural economics," Professor Schultz insisted that research had "reached an impasse" because it had failed either to "provide a basis for guiding" farmers' business decisions or for going "from farm management data . . . to issues of policy."[67] Professor Black found himself "in full accord" with Schultz's "general criticism of much past research." Even those who objected most strenuously mainly took exception only to the degree of the criticism or to Schultz's alternative proposals to make more frequent use of theoretical economic postulates.[68]

This division between research and social action, which became a practical issue in the 1930's, has a parallel in statements commonly

Also see, C. F. Clayton, "Some Types of Economic Research in Relation to Land-Use Planning," *Journal of Farm Economics*, vol. XVI, 2 (April 1934).

[64] M. R. Benedict, "Types of Research Needed as a Basis for Land-Use Planning." Paper read before the American Association for the Advancement of Science, Berkeley, June 19, 1934 (Mimeo.). Also see E. G. Nourse, "Reorienting Research in Agricultural Economics," and especially the discussion of this by F. F. Elliott, *Journal of Farm Economics*, vol. XVI, 2 (April 1934).

[65] J. G. Maddox, "Land Tenure Research in a National Land Policy," *Journal of Farm Economics*, vol. XIX, 1 (Feb. 1937).

[66] C. A. Wiley, "Tenure Problems and Research Needs in the South," *Journal of Farm Economics*, vol. XIX, 1 (Feb. 1937).

[67] T. W. Schultz, "Theory of the Firm and Farm Management Research," *Journal of Farm Economics*, vol. XXI, 3 (Aug. 1939). Also see W. W. Wilcox, "Types of Farming Research and Farm Management," *ibid.*, vol. XX, 2 (May 1938), p. 423; and S. E. Johnson, "Adapting Farm Management Research to New Opportunities," *ibid.*, vol. XXI, 1 (Feb. 1939), pp. 98, 100.

[68] John D. Black, "Dr. Schultz on Farm Management Research," *Journal of Farm Economics*, vol. XXII, 3 (Aug. 1940), p. 570; and "Discussions" by H. C. M. Case, S. W. Warren, G. W. Forster, D. C. Mumford, and R. S. Kifer in *Journal of Farm Economics*, vol. XXII, 1 (Feb. 1940), pp. 111–37.

found in research literature. The 1928 handbook, for example, recognized "pure science" that deals with "a body of generalizations as to the relationships between things without primary concern as to whether the knowing . . . is of any value to anybody." The "distinction between the pure and the applied" is one "that runs across the whole field of science." [69]

In land economics, in 1938, Professors Ely and Wehrwein discussed land economics, which "as science . . . seeks the truth for its own sake," whereas "as an art, it aims to frame constructive land policies." [70] In respect to land tenure research, Professor Duncan, while he insisted that land tenure research improvement was needed because land tenure problems will be "one of the biggest social problems of agriculture for years," at the same time insisted upon an "important and necessary distinction" between a "research problem . . . intended to discover truth" and a "social problem," which has "reference to matters of policy." [71] If the prongs of this dilemma can be brought into a functional relation, perhaps some of the conflicts and confusions that have plagued land economics and rural social science research may be resolved.

The Science of Inquiry

For many years John Dewey has been analyzing the characteristics and functions of science in society, but only relatively recently has his work become available in comprehensive form.[72] This new statement of the development of scientific method has not as yet entered into the operating procedures of social scientists to any great extent. Since Dewey looks upon the field of logic in philosophy as the study of the evolution of scientific method, control over warranted assertions of belief, the more important points in his conclusions demand reference here.

Like practically all historians of science, Dewey sees two major eras of inquiry divided by the period of the Middle Ages.[73] The first era was that of the Greek thinkers — Socrates, Plato, Aristotle and their contemporaries (circa 400–300 B.C.). Their contribution was that of arousing skepticism — intellectual curiosity, a willingness to

[69] *Research Method and Procedure,* p. 403.

[70] R. T. Ely and G. S. Wehrwein, *Land Economics* (Ann Arbor, 1928), p. 2.

[71] O. D. Duncan, "Hypotheses in Land Tenure Research," *Journal of Farm Economics,* vol. XXV, 4, pp. 863–64.

[72] John Dewey, *Logic, The Theory of Inquiry* (New York, 1939). Also see his *The Quest for Certainty* (New York, 1929).

[73] Dewey, *The Quest for Certainty,* Chap. II. J. H. Robinson, *The Mind in the Making* (New York, 1921), Chaps. 8 and 9; B. Ginzburg, "Science," in *Encyclopedia of the Social Sciences,* vol. 13 (New York, 1934).

raise questions. Their method was that of conceptual reasoning. Observation was wholly informal, for operational techniques would have been regarded in their society as "representing an inferior level of activity."[74] Their dominating view was that intellectual certainty was established by perfection in the design or organization of concepts.[75]

Modern science came into existence after the hiatus of the Middle Ages, during which time Greek methods and objectives were retained, but dogmatic faith replaced skepticism. About 1600 A.D. there was a revival of critical inquiry by a few courageous individuals, and their activities are acknowledged as the beginning of science as we know it today.

Of the great names in this early movement, one of the most important is that of Francis Bacon, who in his *Novum Organum* sounded an epochal call for a search after new facts to advance knowledge. It is from this important challenge to the moribund philosophizing of the Dark Ages that collection of facts has come to be regarded by many as the critical mark of modern science and as the antithesis of the Greek position. Karl Pearson's presentation clearly indicates this influence.

No doubt the Baconian view is in sharp contrast to the Greek in respect to the materials of inquiry. But in Isaac Newton's *Principia,* a work contemporary with Newton's, science retained the Aristotelian aim of finding certainty by the discovery of an orderly arrangement of universal truths. This objective is also clearly seen in more modern form in the work of Karl Pearson.[76]

In view of this background, current concepts of science are likely to be based on these elements: (1) the *skepticism* of the Greeks, (2) the Baconian emphasis on *fact collection,* (3) the Aristotelian-Newtonian aim of finding great, brief *summaries of orderliness,* (4) the Pearsonian techniques of stating the summary in terms of *probability formulas.*

John Dewey, in his analysis of the history of scientific method, finds an important oversight in this formulation. The great omission is that of *experimentation* as the crucial mark of modern science. The difference between ancient science and modern science turns on the use of experimental operations. And as this new method has been used,

[74] Dewey, *The Quest for Certainty, passim;* Robinson, *The Mind in the Making;* Ginzburg, "Science."

[75] "What the Greeks sought was in the nature of a theoretical, contemplative Weltanschauung." Ginzburg, "Science," p. 596.

[76] Karl Pearson, *Grammar of Science,* pp. 87–88; Dewey, *The Quest for Certainty,* p. 142; Dewey, *Logic, The Theory of Inquiry,* p. 127.

conceptualizing has been given a new place in the hierarchy. There has been a drift away from the idea of finding brief universal generalizations as the goal of science.[77]

Galileo introduced a new method of inquiry in that his actions were "deliberate in the scientific sense of being performed with a view to fulfilling an intellectual end . . . for the purpose of testing a set of ideas." And it is only "after some three hundred years of continuous use and development [that] the new method of scientific inquiry has succeeded in finally destroying Aristotelianism in the technical field of the most important natural sciences." But this has not yet come to pass "in our current culture generally and in our social 'sciences.' "[78]

Since no one doubts the experimental character of modern natural science, an understanding of modern science requires an analysis of the place and function of experimentation. In view of the fact that interpretations of modern scientific method often center on the handling of quantitative data derived from experimental operations rather than on the procedure of experimentation itself, attention to the latter may clarify certain perplexing issues which have been shown to exist.

A most important characteristic of experimentation is that it is in itself a "functional integration of theory and practice," of ideas and actions, of knowing and doing.[79] The experimenter does things with the elements in which he is interested: he assembles, arranges, and rearranges his subject matter; he observes these materials in action and overtime. What he does and how he does it is determined by his ideas and concepts, and what he sees and does continually affects his ideas and concepts. He also takes records and makes symbols of what he sees and does; and these he arranges and rearranges in order to furnish further clues and to revise his ideas as to how to do things differently in his physical experiments. He keeps on doing these things until he has found the means which, when instituted in the actual experiment, produce a postulated outcome.

In other words, in experimentation there is no rejection of ideas or conceptual reasoning, but such activities serve the function of directing action; and the resulting actions serve, both directly and indirectly, to modify the ideas which are held by the scientist. "The test

[77] Ginzburg, "Science," p. 602.

[78] Joseph Ratner, "Introduction to John Dewey's Philosophy," in Dewey, *Intelligence in the Modern World* (New York, 1939), pp. 68–70. See also H. S. Fries, "Science, Ethics, and Democracy," *Journal of Social Philosophy*, vol. VI, 4 (July 1941), p. 311.

[79] Ratner, "John Dewey's Philosophy," in Dewey, *Intelligence in the Modern World*, p. 114.

of ideas . . . is found in the consequences of the acts to which the ideas lead."[80] For "experience when it is experimental does not signify the absence of large and far-reaching ideas and purposes. It is dependent upon them at every point. But it generates them within its own procedures and tests them by its own operations."[81]

This inseparable relation in experimental inquiry between the ideas of the scientist and the actions in the laboratory has its counterpart in the relation between science and experience. "Scientific inquiry always starts from things of the environment experienced in our everyday life."[82] Where there is doubt or conflict in respect to that experience, there is a starting point for a line of scientific inquiry. The function of science is to determine ways of acting that will bring activities to a stated consequence. Science is a continuing process of problem-solving in order to give man better control over his experience.

The problems of inquiry arise from problems in experience. In a given situation, doubts, confusions, or conflicts arise as to the outcome of an event. The question is this: What means, if instituted, will produce what consequence? The question is posed in the setting of a problem situation in experience.[83] In actual experience, various elements act and react on each other, within an environment and over a period of time.

The scientist works with suggestions, ideas, and concepts that may be formally or informally obtained from experience in solving previously raised problems,[84] formulating the problem for inquiry by noting certain elements which seem to be strategic in the problem event. These elements are then put into interaction in the laboratory and the outcome observed. The elements may be rearranged, operations undertaken, and the outcome observed again. "The ground and criterion of the execution of this work of emphasis, selection, and arrangement is to delimit the problem in such a way that . . . material may be provided [from experience] with which to test the ideas that represent possible modes of solution. Symbols, defining terms,

[80] Dewey, *The Quest for Certainty*, p. 136.

[81] *Ibid.*, p. 138.

[82] *Ibid.*, p. 103. Cf. Ginzburg, "It is a relic of an outworn meta-physical conception of science to suppose that science can be carried on . . . without reference to practical viewpoints and interests." "Science," p. 601. See also Dewey, *The Quest for Certainty*, pp. 100–2.

[83] Dewey, *Logic*, pp. 106–8.

[84] The order, which is often taken to be the chief characteristic of science, refers but to the desirability of neatness and consistency in arranging the conceptual tools and results of experience gained in the solution of previous problems so that they may be most efficiently reached and cross-checked as new problems come up for attention. See Ratner, "John Dewey's Philosophy," in Dewey, *Intelligence in the Modern World*, p. 110.

and propositions, are necessarily required in order to retain and carry forward both [ideas and factual materials] in order that they may serve their proper functions in control of inquiry."[85]

A hypothesis is constructed out of the suggestions and ideas by which the problem has been tentatively formulated. The beginning hypothesis is worked into the form of a proposition in which an "if" clause states a possible something-to-do and a "then" clause postulates the consequences of such action. The hypothesis directs the investigation. It is the basis for tentative selection of facts as evidence from all the facts in the situation.[86] The formulation of the problem and the hypothesis, always tentative, are subjected to expansion, revision, modification, and refinement until the hypothesis proposes means which, when instituted, do result in the stated consequences.

Deduction and induction then are seen not as alternative steps or methods of inquiry, but as referring to the techniques of (1) developing directing concepts and (2) preparing the facts, which two functions are inseparable in experimental operations. "As far as *processes* of inquiry are concerned, there is no difference between induction and deduction."[87]

The procedures needed to handle and detect relevant facts must be determined by the problem at hand, just as laboratory apparatus and devices are developed to suit the type of materials involved in the problem. There will be a priori suggestions for possible techniques from previous experience in inquiry in respect to both conceptual and factual materials. But there is a priori no one procedure, inasmuch as each problem constitutes a challenge to devise new ways and means of formulating directing concepts, handling factual evidence, and instituting controls. Such inventions may, in turn, cast new light on methods previously determined for resolving other problems and may even result in wholly new formulations of problems previously studied. In all cases, "familiarity with material, sagacity in discrimination, acuteness in detection of leads or clews, persistence and thoroughness in following them through, cherishing and developing suggestions that arise" are required of the scientific investigator.[88]

[85] Dewey, *Logic*, pp. 117–18.
[86] *Ibid.*, pp. 112, 272–73, 498, 503–11.
[87] *Ibid.*, p. 484 and Chap. XXI. A differently formulated but not inconsistent resolution is presented in Kaufmann, *The Social Sciences.*
[88] *Ibid.*, p. 485. See also pp. 205, 210, 484, 499–503, and Chap. XXIII. Note also: "It is one thing to speak of the 'scientific attitude,' . . . It is another thing to speak of '*the* scientific method.' Because no such animal exists. A scientific method develops and changes as the particular science whose method it is develops and grows. Where does it come from? From the only place it can come from; namely, from the concrete subject-matter of that science." H. S. Fries, "On the Meaning of Intelligence," *Educational Administration and Supervision* (Jan. 1936), p. 39.

In view of the persistence of the idea that the goal of science is the formulation of generalizations, what is the place of generalization? Put another way, what is the relation between generalization and problem-solving in the sense of instituting means to attain stated consequences? In scientific inquiry, generalizations may be made of conceptual materials or of factual materials; both forms of generalizations are used in inquiry. However, they are sought not as ends but as suggestions for possible ways to resolve the problem under investigation.

In experience, in respect to both place and time of occurrence, an event is a unique qualitative situation. Problems arise in such a setting, and problem solutions are finally tested in such a setting.[89] In contrast with this qualitative and sequential nature of the origin of problems and the final test of their solution, the elements of a law or generalization are not in such a relation. The traits or elements of a generalization are "*logically*, not temporally, conjoined. They are selected and ordered . . . into a definite set of interactions."[90]

Since the final test of inquiry is in an event, how can the "laws of science," on which so much emphasis is often placed, be judged? They must be regarded functionally; that is, they are important not as final grounds in inquiry, but because they suggest possible sequences within events. They are "instrumentalities in determining, through operations they prescribe and direct, the ordered sequence into which gross qualitative events are resolved."[91]

Social Research

Does this view of the character and function of science offer a possible basis for a mode of resolution of the doubts and confusions that appear in discussions and evaluations of social science method in general, and rural social science and land economics in particular?

Most obviously, it offers a framework of scientific method that is comprehensive enough to include a place for issues that are not faced in most formulations of science that have otherwise been offered.

Whereas Dewey's presentation of modern scientific method (1)

[89] When the elements in a problem situation become "integral constituents of one and the same continuous occurrence," then this "determination . . . constitutes the resolved individual qualitative situation which is the final, or terminal, conclusion." "Science resolves the gross qualitative events . . . into a set of interactions, each one of which . . . is capable of uniting with others to form a continuous coexistential-sequential whole with gaps and interruptions." Dewey, *Logic*, pp. 445–46.

[90] *Ibid.*, p. 454.

[91] *Ibid.*, pp. 454–56. They are "means, through the media respectively of operations of reasoning (discourse) and of observation, for determining existential . . . connection of concrete materials in such a way that the latter constitute a coherent individualized situation."

starts with problems arising out of experience, (2) includes the formulation of problems for controlled inquiry, (3) covers experimental operation, (4) shows the function of conceptual and factual generalizations, and (5) finds the terminal test of inquiry in experience, the predominant concepts of scientific method concentrate only on the establishment of factual generalizations and either overlook or make but vague references to the other aspects of modern experimental science.

It should be noted that Dewey's analysis of "*modern* science as exemplar of what knowledge is, and the *method* of *modern* science as standard of the method of knowing"[92] is based on analysis of physical and biological, not social, science. To Dewey, philosophy is a social science that deals with human beliefs, including those arrived at by scientists, and their relationship to society.[93] Also, Dewey has been interested in stimulating the wider use of scientific methods of knowing to social affairs.[94] But because of the advanced stages of natural science, it is used by Dewey as the exemplar of the modern method of knowing. In his *Logic,* therefore, there is one special chapter out of twenty-five that is entitled "Social Inquiry," and this is inserted because "The very backwardness of social inquiry may serve . . . to test the general logical conceptions that have been reached."[95]

The main points of Dewey's chapter on "Social Inquiry" will be given briefly here before an attempt is made to make a further specific clarification of the issues. The following outline summarizes Dewey's contribution in Chapter XXIV of his *Logic.*

I. Comparison of research problems in natural and social science reveals that although social research has some special problems that make it difficult, these do not destroy the possibility of developing social science.

II. In social inquiry, theory and practice are pushed widely apart by the tendency either (1) to assume a problem instead of formulating it analytically, or (2) to assume that all that needs to be done is to assemble and generalize facts.

III. Problems for social inquiry must grow out of social troubles, must have their subject matter determined by the elements that can be used as means in solving the problem, and must be related to a hypothesis which is a plan for resolving the actual social difficulty.

[92] Ratner, "John Dewey's Philosophy," in Dewey, *Intelligence in the Modern World,* p. 57.

[93] Dewey, "Philosophy," in Gee, ed., *Research in the Social Sciences,* p. 256.

[94] Ratner, "John Dewey's Philosophy," pp. 69–73.

[95] Dewey, *Logic,* p. 487.

IV. Facts in social inquiry are either obstacles or facilities for changing social experience; and all social facts are necessarily a part of of a historical process.

V. Concepts or theories in social inquiry often tend to be taken as final truths rather than as tentative suggestions for operations; they often are too loosely formulated; and their usefulness is decreased by the compartmentalization of social science fields. Finally, social researchers should take advantage of such social experiments as are put into operation by making discriminative observation of consequences, for social facts can be understood only in their connection with consequences that follow from an instituted plan of action.

Can we go further than this analysis in pointing up a possible method of resolution of the confusions noted? More specifically, what is the direct import of the experimental nature of modern science for social research? Granted that social phenomena present special and complex practical difficulties for science, if these difficulties are to be an "intellectual stimulus and challenge to further application"[96] rather than an excuse for inaction, how to proceed?

As has been pointed out, modern experimental science has too often been interpreted narrowly in terms of some of its techniques for handling data.[97] But there may be somewhat similar difficulties in stressing the adjective *experimental.* Experiment is commonly interpreted as the literal transplanting of elements from experience to a physically controlled laboratory. This interpretation leads to the "idea that because social phenomena do not permit the controlled variation of sets of conditions in a one-by-one series of operations, therefore, the experimental method has no application at all." This notion "stands in the way of taking advantage of the experimental method to the extent that is practicable."[98]

On the other hand, it may be that some statements concerning the experimental method have been so stated as to discourage greater attention to the possible role of experiment in modern social science. Thus, Dewey at one point has stated that "if we want something to which the name 'social science' may be given, there is only one way to go about it, namely *by entering upon* the path of social planning

[96] *Ibid.*, p. 491.

[97] Cf. *ibid.*, p. 498: "The assumption that social inquiry is scientific if proper techniques of observation and record (preferably statistical) are employed (the standard of propriety being set by borrowing from *techniques* used in physical science) . . . fails to observe the logical conditions which in physical science give the techniques of observing and measuring their standing and force."

[98] *Ibid.*, p. 509.

and control." For, "The building up of social science . . . is dependent upon putting social planning into effect."[99]

Although Dewey in the same place insists that he is "not arguing for the desirability of social planning and control," still the choice he leaves seems to be that of foregoing the use of scientific method or of deliberately instituting new general controls over human affairs. Lynd's arguments also create the same impression.[100]

It is possible, however, to suggest an interpretation that is not inconsistent with the basic methodological position and that avoids these implications. All human activity can be looked upon as the institution of action for some end in view; there is no reason to suppose that only after the inauguration of some particular arrangements will people generally begin to have plans of action. If control is regarded as the use of foresight of possible consequences and the adoption of modes of behavior to try to achieve certain aims, then, within the limits of their intelligence, people do attempt some kind of control over their experience. Social science, then, does not need to stand and wait for the establishment of a new system of government or economy; it can be put to work continuously as a means of helping individuals and groups to use foresight and to relate their activities more successfully to outcome.

As long as people have purposes and practices or ways of doing things (and there seems to be no doubt that these always exist), the necessary elements for social science are present. The available instruments for directing certain activities toward certain ends may increase, or there may be increasing complexities in aligning actions and purposes; but always there is need to use intelligence better to harmonize action and consequences. It is for this purpose that we seek to improve knowledge. With this point in mind, we return to the implications of the experimental character of modern science for social research.

Writers like Karl Pearson overlook the experimental character of modern science and concentrate on its methods of handling data in discussing the application of science to social problems. Others, like Lundberg and Mitchell, offer the statistical method as social science's substitute for experiment.[101] Others, like Elmer, Odum, and the 1928 committee, consider the experimental method one of several alternatives that the research worker in social science might choose, and fail

[99] John Dewey, "Social Science and Social Control," *New Republic* (July 29, 1931). Also in Dewey, *Intelligence in the Modern World*, pp. 951, 954. Italics inserted.

[100] See footnote 60 above. Also, Lynd, *Knowledge for What?*, *passim*.

[101] Lundberg, *Social Research*, pp. 48–53; W. C. Mitchell, "Quantitative Analysis in Economic Theory," *American Economic Review*, vol. XV, 1 (March 1925); and F. S. Chapin, "The Experimental Approach," *Social Forces*, vol. XI, 2 (Dec. 1932).

to give it any general function comparable to its place in other sciences.[102]

How, then, can we recognize more specifically the import of experimentation for social inquiry? Two suggestions may be found helpful in formulating an answer.

The first implication of the experimental character of modern science for social science is that its problems are set in terms of actions and outcome of actions. The function of social science is that of resolving problems that arise where there is confusion regarding the unity between what is done and the consequences of the doing. When a situation is so altered that the action taken results in the end-in-view, the problem is resolved — the situation is unified.

This conception of problems is borne out literally by the experimental actions and operations of modern experimental science. It can be explicitly applied in social inquiry. A study of a community or a survey of land utilization is not an attack on a problem of experience; whether it may be helpful to the solution of a problem cannot be known until the problem is at least tentatively formulated. Similarly, a study of the relation between land use and soil type or a study of factors affecting forestry is not a formulated problem of experience. It may be necessary to the solution of some existent problem, but again its relevance is unknown as it stands. A problem must be rooted in an ongoing event in which there is confusion respecting action consequences. Thus a study of what can be done about landlord-tenant relations so that the farms of a community will be conserved is at least a rough statement of a problem in experience. If in an inquiry into such a problem it becomes necessary to study the relationship between tenancy and soil type, then this question, a problem in inquiry, has a known root in a problem in experience where its final test is to be found. Without the first formulation of the original problem, the work of inquiry is aimless and may be sterile.

From the tentative formulation of a problem in experience, it is possible to construct a hypothesis that will be practically useful — that is, one that will direct the work of inquiry. The hypothesis postulates, in respect to the problem situation, that if (and only if) thus and so is done (with whatever elaborations may be necessary) then thus and so results.[103]

[102] Elmer, *Social Research*, Chap. XIII; Odum and Jocher, *Introduction to Social Research*, Chap. XVII; *Research Method and Procedure*, pp. 314–22.

[103] It should be pointed out that because simplified illustrations are used in methodological discussions, there is a tendency to regard hypotheses (either beginning or concluding ones) as limited to a single sentence form. Of course there is no necessity for such brevity in the concept of hypothesis. In fact, a progressive inquiry may start with

Clearly, neither the elements in a situation or the consequences are singular; they are multiple. The big task is to sort out the irrelevant, the relevant but complementary, and the strategic means (actions to be taken) and ends (consequences or outcomes). The hypothesis is a tentative formulation in this regard. It is a working hypothesis because it suggests the means to be tried and the outcomes to be watched. It is tentative because it is continuously revised in the process of inquiry. Without some hypothesis, the inquiry will be at least inefficient; without a flexible, tentative hypothesis it may be dangerous.

The second implication of the experimental character of modern science has to do with arrangement of the evidence. What does it mean, practically, for social science to use procedures which, if they are not the exact duplicate of the experimental laboratory method of physical science, are at least functionally similar in respect to handling evidence? How can the strategic characteristics of the experimental method in this regard be preserved?

One of the characteristics of the experimental method in respect to evidence is that the nature scientist puts the elements of his problem into action under controlled conditions. He conducts operations in rooms with controlled light or temperature, in devices that may hold certain elements constant and that may entirely eliminate factors that, at the moment, are considered irrelevant, or in fields with certain known soil type, and so forth. Yet the scientist does not get the same degree of concrete control in every problem; the control depends upon the nature of his problem and upon the stage of research on it. Sometimes he has to carry himself and his equipment to distant locations because he cannot put the critical elements under control in his own laboratory or because for other reasons the observations can best be made at those geographic spots. Even at such places, the scientist may not be able to control physically the phenomena which are relevant, but in such a case he has to take special precautions in watching for the influence of factors which may be thought irrelevant but which might be strategic.

It is but an extension of these adjustments, which are imposed by the nature of the problem in other sciences, to state that the elements in a social problem can seldom if ever be transplanted and set in motion under conditions rigorously determined by the scientist

a rather vague but very brief hypothesis and conclude with the hypothesis restated at great length but with specificity. Beginning hypotheses are not simply to be adjudged true or false; rather they are to be modified until they present a way of unifying actions and ends.

specifically to satisfy a question in inquiry. Conversely, it is only an extension of the same considerations in physical science which make it necessary for the social scientist to make his observations by taking advantage of such instances as appear to have a bearing on the problem he is studying wherever and whenever they occur. In other words, he must be on the lookout for relevant experiments wherever they occur.

These conditions impose special problems for the social scientist, especially, of course, that of guarding against oversight of elements that are more important than the hypothesis suggests. But they are conditions to be faced and overcome, rather than excuses for removing social inquiry from the influence of modern scientific method. Some of these matters are involved in what Professor Sorokin calls "logico-meaningful" analysis in which observations may occur "at quite different periods, in quite different places, and only once or a few times."[104] Yet the justification for such a technique rests not on any "higher" purpose of social inquiry nor upon any functional difference between the science in the two fields, but merely upon the impossibility of identical operational techniques — which impossibility exists among the various natural sciences as well as between natural and social sciences in general.

As long as people as individuals or groups do certain things to attain certain consequences, then all activity is in some degree an experimental operation, whether or not there is any correspondence between the things done and the consequences sought. A number of these activities may seem to offer potential evidential material for a given problem. The relevant environing conditions may not be the same among these experiences, but the variations in these events, insofar as they are relevant, offer situations functionally similar to those found in the operations in a scientific laboratory.[105]

In short, human experience may be viewed in terms of means-consequences patterns, and these patterns of experience can be regarded as experiments. The difference between this position and the work of the physical scientist is one of degree and technique. The social scientist often exaggerates in his own mind the extent of this difference; for the degree of artificial control obtained in any experiment depends upon the nature of the problem, the type of subject matter, the state of knowledge, and the practical limitations imposed on the project.

[104] Sorokin, "Culture Integration," *Rural Sociology*, vol. I, 2 and 3 (June and Sept. 1936), p. 350.

[105] F. S. Chapin calls these "ex post facto experiments." See Ernest Greenwood, *Experimental Sociology* (New York, 1945), especially Chaps. VIII, IX.

Secondly, however, an even more important point is that laboratory experiments have a definite comparability to the existing situations out of which social problems arise and in which they are finally resolved. This comparability lies in (1) the existence of interactivity among the elements characteristic of both a social event and a laboratory experiment, and (2) the sequential, temporal, or process also characteristic of both.[106]

The scientist calculates relationships between variables; that is, he manipulates factual observations drawn at different points from his experimental operations. Nevertheless, in the laboratory he has these elements interacting in continuing relations organically and over time.[107] The data which he records from various points in his experimental observations are worked into factual generalizations, and very often it is these relationships which are printed as evidence and which impress the outsider. But the experimental scientist has watched the integral functioning process from which these observations have been drawn. Their interactions and sequential *relations* have been under view in the laboratory operations.

Under these circumstances, it becomes all too easy to look at the conceptual and factual generalizations of relationships which the scientist produces. It is even easier to forget the conditions under which the underlying data were drawn and also that the generalizations are used by the scientist as suggestions on how to set these elements working on one another and over a period of time in actual laboratory experience.

Laboratory operations, in which there is a chain of interactions among the elements of the problem, provide the next-to-final test of the hypothesis (proposed mode of activity for postulated consequences) as then stated before its terminal test in experience, in which again there is a chain of interactions among the elements and with their environment.

To come as close as possible in social science to a functional parallel to this experimental method of modern science, it is here emphasized that the next-to-final test of social science requires that the evidence must reveal the chain of actions actually existing among the relevant elements of the problem. Such a chain of action can take place only through acting human units.

[106] Cf. E. C. Lindeman's concepts of impulsion, circumjacence, interaction, and emergence in *Dynamic Social Research*, Pt. 2. Circumjacence and interaction may be regarded as interaction of the elements in the problem (1) with their environment and (2) among themselves. Impulsion and emergence may be regarded as process of these interacting elements from (1) institution of action to (2) consequence.

[107] Dewey, *Logic*, pp. 400, 468–69.

It is, therefore, the thesis of this review of rural land economics research that confusions, conflicts, and doubts have been accentuated by failure to see the essential characteristic of modern science — of the experimental method of knowing. Social scientists have not looked into the entire process of science to see where it has made the greatest advances. They have not generally noted the source and solution of problems in means-consequence relations, the importance of problem formulation, the interpretation of hypotheses as tentative modes of solution, the functional correspondence of ideas and operations, the suggestive role of generalizations, and the critical characteristic of experimental testing as that of producing a process of interaction among the elements of the problem.

Instead of working on the admittedly large problems that these aspects of modern scientific method pose for social inquiry, social scientists have concentrated on only one or two of these segments, or they have gone off in other directions — some as far as to renounce the possibility of finding scientific method useful in the solution of human problems. If social research problems were more clearly formulated in terms of ongoing experience, if hypotheses of proposed means for postulated consequences were used in directing the analysis (the selection of relevant facts from all facts), if factual and conceptual relationships were seen as only suggestive of possible sequential relations, and if the conclusions were grounded in observed interactions in experience, then rural land economics research would be more reliable and more useful.

Research Procedures

In the literature of rural and general social science there is considerable vagueness and contradiction among the many attempts to classify and distinguish various research methods, procedures, and techniques. To attempt to codify these presentations would be a tremendous task and probably would not be worth the effort.

In the 1928 handbook, for example, there are stated to be five research methods, but research is also classified on the basis of three approaches. Again, there has been no clarification in agricultural economics of the methodological cleavage between the procedures of those who followed in the Warren farm management tradition and those who followed in the Taylor tradition. Many of the latter group became devotees, during the 1920's, of the statistical method and were most influential in the preparation of the 1928 handbook. Yet in 1931 Professor Warren defended the school of thinking which he led by saying that the agronomists who developed farm management "carried over their scientific method at once into their work." "It may not always have been good science, but it was nevertheless the scientific point of view," Warren asserted, and he went on to observe that "Since the scientific method was used, it was from the start expected that statistical equipment and clerical help would be involved." [108]

As early as 1911 Taylor described five methods of agricultural economics research: the historical, geographical, statistical, accounting, and experimental.[109] But Taylor thought of the first three as ways of using data collected by the census and by market reporting agencies, and the last two as designed for individual problems in specific instances.

Because Taylor had no provision for ways of collecting data from relatively large numbers of farms—that is, he had no formal means for studies intermediate in scale between total census data and individual cases—Warren's survey method of amassing data came to be regarded as another and a different research method.[110] Taylor did, in fact, publish a research bulletin in which he used as evidence the results of interviews with numerous farmers,[111] but since he did not use any of this information quantitatively, it was an informal method, one sometimes referred to orally as the "Wisconsin essay" method.

Taylor's historical and geographical methods were mainly distinguishable by the fact that mass quantitative data were arranged chronologically in the one method and on maps in the other. The word *geographical* was therefore inappropriate to Galpin's work in which information from a number of direct sources taken in the field was arranged on maps; this was later referred to as ecological analysis by the sociologists.[112]

In general, social research methods, procedures, and techniques have been differentiated on the basis of the amount of evidence available, the subject matter of the data, the type of problem, the ways of collecting evidence, arranging it, or presenting it to others, or the professional training of the investigator. Some presentations have included several of these possible bases of classification in one system. Thus Lindeman in 1925 recognized historical, analogical, logical, and sta-

[108] G. F. Warren, "The Origin and Development of Farm Economics in the United States," *Journal of Farm Economics*, vol. XIV, 1 (Jan. 1932).

[109] H. C. Taylor, *The Place of Economics in Agricultural Education and Research* (Wisconsin Agr. Expt. Sta. Res. Bull. 16, 1911).

[110] G. F. Warren, *Agricultural Surveys* (New York [Cornell] Agr. Expt. Sta. Bull. 344, 1914).

[111] H. C. Taylor, *Methods of Renting Farm Land in Wisconsin* (Wisconsin Agr. Expt. Sta. Bull. 344, 1914).

[112] J. C. Galpin, *The Social Anatomy of an Agricultural Community* (Wisconsin Agr. Expt. Sta. Res. Bull. 34, 1915); and *Rural Sociological Research in the United States*, p. 70.

tistical methods and urged the addition and integration of the psychological; then, in 1933, he listed interviewing, observing, case analysis, and charting techniques and statistical devices.[113] As previously noted, the 1928 committee recognized an experimental method whereas some others do not. Some students place the case and statistical methods at opposite poles while others regard the latter as a mere extension in quantity of the former. At one point, Taylor has listed the analytical method as additional to the historical, geographical, statistical, accounting, and experimental—a reference comparable to those to a special logical method of research.[114]

In this section, instead of attempting to establish a system of research method or procedure categories, reference will be made to some of the terms currently used in respect to research method, insofar as such reference can be used to clarify the hypothesis above presented.

AN OUTLINE OF INQUIRY

In order to focus this discussion, the suggestions developed in this chapter may be presented in brief outline. This outline covers a full research inquiry. It is important, therefore, to recognize that *research project* is a relative term. Because of personal interests and aptitudes, or because of work being done elsewhere, or because of administrative limitations, a research project may embrace but a minute part of this outline. The important thing, however, is that a project that is but a partial segment of a full inquiry, to be of maximum value, should be so structured that it is clear how it fits into the full inquiry. Criticism of a partial study must be bound by the limited purposes of such work, but that does not eliminate the importance of making clear what part it is intended to play in the full task.

Also it should be borne in mind that there is an inherent danger in an outline of this sort—the tendency to regard its parts as closed compartments. In any inquiry any phase of this outline may be in action at any time. Also there is gradation between its parts, not all-or-none partitions.

I. *Problematic Situation:* Doubt as to relation of action and outcome in experience.

II. *Formulation of Problem:* Tentative designation of actions that are possibly strategic to the consequences in the situation.

   1. Function: To eliminate the more obviously irrelevant elements in the total situation.

[113] Lindeman, *Social Discovery* and *Dynamic Social Research.*

[114] H. C. Taylor, "Agricultural Economics from 1908 to 1911" (unpub. ms., May 17, 1941), p. 45; Lindeman, *Social Discovery,* Chap. III; and footnote 39 above.

2. Method: Use available facts and concepts, results of prior inquiries, and hunches, ideas, and suggestions; as inquiry proceeds, use evidence produced in current study.
3. Status: Tentative; subject to revision at any time.

III. *Hypothesis:* Tentatively proposed statement of what actions result in postulated consequences.

1. Function: To direct search for relevant evidence out of welter of facts in the total situation.
2. Method: Use available facts and concepts, results of prior inquiries, and hunches, ideas, and suggestions; as inquiry proceeds, use evidence produced in current study.
3. Status: Tentative; continually revised and elaborated all during the processing of evidence.

IV. *Processing Evidence:* Constructing or finding situations in experience in which postulated actions and consequences form integrated events.

A. Evidence of *relationships*

1. Function: Instrumental, to suggest possible relations among elements in the problem.
2. Method: Use various known techniques and develop new ways to construct generalizations concerning the conjunction of particulars among the observations of actions and consequences.
3. Status: May range from impotent to highly suggestive in respect to relations.

B. Evidence of *relations*

1. Function: To test the interaction and sequence of actions and consequence of actions, and consequences with actual experiences.
2. Method: In physical science, experimental operations provide the qualities of interaction and process; in social science various techniques are used to expose interaction and process in the continuum of people's experience.
3. Status: Has greater testing force the more gaps are eliminated in the sequential course of observed experiences.

V. *Terminal Test:* Purposive action is instituted and is consonant with consequences.

1. Function: Actually to solve the problem.
2. Method: The application of intelligence to direct experience.

3. Status: Scientific inquiry has no ultimate stopping point since experience is continuous and ends-in-view are altered as a result of every experience.

EXPERIMENTAL METHOD

Modern science is experimental. If, in social science, the investigator seldom literally institutes means to observe consequences, he performs the same function when he directly or indirectly observes the chain of interactions from purposive action to outcome in the experiences of others, insofar as the experiences appear relevant to the problem with which he is concerned. If a given research project is in itself but a partial attack on a study, it must be somewhere tied in to a broader experimental inquiry to make it purposeful research.

HISTORICAL METHOD

As modern science takes hold in social affairs, historical method becomes less distinguishable. Modern science "introduces history into everything . . . and nothing is intelligible without some knowledge of its past."[115]

Since human experience is a temporal, moving process, all evidence with testing force is historical. Historical evidence is not only relevant to the dead past. Though no present event can be identically similar to any other past (or present) event in all respects, past events provide the only possible experimental evidence—reliable evidence if careful account is taken of the factors which distinguish one circumstance from another and if the relevant and irrelevant elements are differentiated. Furthermore, history does not start and stop in the past; past experience reaches up to become the moving present in which problems are felt and solutions are applied.

If carefully selected strategic elements in the action and consequences of a historical event correspond to those of the current problem, and if that event is analyzed so as to show interaction and process in the experience of people, it has testing value. But if historical materials are but dated facts and only show relationships among these particulars, then they serve an instrumental purpose.[116]

The high respects paid to the scientific utility of historical research, even by those who give weight and emphasis to other methods, may be owing to the fact that historians have developed high standards of

[115] Julian Huxley, "The Biologist Looks at Man," *Fortune,* vol. XXVI, 6 (Dec. 1942), p. 139.

[116] Thus, Murray Kane in criticism of certain of Turner's work in American history: "What results primarily is not a historical interpretation of statistics but a statistical interpretation of history." *Mississippi Valley Historical Review* (Dec. 1940).

reliability for evidence of past events and often produce results that do show closely-knit sequential experiences of individuals and groups. This method, therefore, has often supplied a type of evidence that has strong testing force in social inquiry.

If historical research is criticized, it may be because of failure to engage in it without reference to a problem or a hypothesis.[117] In history for history's sake there is no directing guide as to the relevancy of evidence, no basis for judging its connection with problem-solving. Its only standard is completeness. The results, then, may be interesting and entertaining; they may sometimes be somehow helpful to some research, but they are not examples of social science inquiry.

CASE METHOD

The case method has been fought over bitterly in the wars of social science methodology. Even its strongest foes, however, favor it with high praise. Yet it is nearly impossible to find a coherent exposition of it. When is a case a "case," and when is it a sample or a unit of observation within a sample? Can statistical devices be used to study the constituents of a case, or do several cases compose a statistical method? Is not a historical study of an event a case study? These questions may be looked at in terms of the above functional outline of research.

If a case study is, as commonly interpreted, an intensive study of everything that bears on a given unit, then there is no formulated problem and no hypothesis, and the work is not a scientific inquiry.

If a case as evidential material is but a sample area or a sample group of units of observation, it may be analyzed in such a way as to offer evidence of relationships which have instrumental or suggestive usefulness.

If a case is an acting unit and if the interactions and sequences in its experience are preserved within the unit, then it has strong testing force. Insofar as the interaction and sequential gaps among the facts of the unit of action's experience are closed in a case study, and as far as these facts are relevant to the experienced problem under study, to that extent a case study has the quality of testing relations in the only place where they have meaning. In this form the case method can prepare evidence that carries exceedingly great weight as a test.

It will but rarely be true that analysis of a single case (in the sense of a single unit of action) will suffice for a full inquiry. There have to be as many cases as there are combinations of strategic means-ends factors for a full analysis of a problem. But this number is required

[117] Cf. Dewey, *Logic*, Chap. XII.

not because of any need to make a lump sum of large size, but rather so that there will be enough separately distinguished experiments to reveal the processes of the various possible means-end events. A larger number of cases will be mainly useful as checks on the adequacy of the determination of elements as strategic, complementary, or irrelevant. But if the use of a larger number of cases throws the analysis over to a study of relationships between observations, taken out of their sequential relation, then it is important to remember that the function of such analysis is changed to that of suggesting possible relations which still must be found to exist in the experience of each case if it is to stand as testing evidence.

QUALITATIVE DESCRIPTION

Like the usual discussion of case or historical methods, qualitative description also receives high praise even while it is relegated to inferior status. But again, it is its function that is important rather than its status in a hierarchy of methods.

If the physical scientist uses qualitative description less than the social scientist, it is not because he is more scientific. It is simply that since he and other scientists can duplicate the final experimental test, they can provide themselves with a chance to see the problem-solution in process and do not need to read or write an elaborate description of it. It is the activity of the physical scientist in the testing laboratory that parallels the work of the social scientist when he spells out in written detail the unfolding story of the persons or groups whose experiences comprise his experiments. In the performance of this function, it is important for the social scientist to be as careful as possible in his choice of words and in his presentation in order that other social investigators may clearly comprehend the processes that he has uncovered without having to search out the same evidence themselves. It is just as important for the social scientist accurately and carefully to describe the experience of the people that is his experiment as it is for the physical scientist to use care in setting up his laboratory equipment and performing his operations. The function of experiment is the same in both instances, and its testing force is equally great.

The materials of social science may not always be as readily subject to quantification as are the materials of physical science, but the materials of social research probably are equally susceptible to other forms of symbolic recording. This ability to symbolize is not what requires more qualitative description in social science than in other science. What does make the difference is in the one case the ready ability actually to institute actions and to observe experimental process, and in the other the necessity of using verbal descriptions to describe it.

ANALOGICAL METHOD

No method is more universally exercised than analogy. It is invidiously compared with other more scientific methods and is often judged on the assumption that only incompetents would make use of it. If it is more inexpertly used than other methods of learning from experience, it may be owing to its generally more frequent use rather than to any inherent limitation of the method. For it may also be said that all experience and science are analogical since any purposive action controlled by intelligence is controlled by what has been learned from past experience in nearly comparable circumstances.

Thus, if analogy is looked upon in terms of its function, it is seen to be not some special method that has to be rated against other methods, but a part of all inquiry. A careless analogy simply is inquiry of poor quality; careful analogy is inherent in research.

If the concepts of research here developed are applied to analogy, it is clear that analogy may or may not provide testing evidence. An unrefined use of analogy may be very helpful in formulating problems or hypotheses. Analogy that only compares particulars among cases may be instrumental in suggesting further tests. But analogy in the sense of comparing relations within a case with relations within other cases has high testing force, and the degree of its testing force is the degree of unity in the internal relations and in the sequential experience of the cases. In this latter form, analogy is of course very widely used in the most advanced sciences and is the test immediately preceding the terminal tests in actual experience.[118]

LOGICAL METHOD

As in the case of other methods, references to the logical method are seldom given functionally and thus are likely to be confusing and apparently contradictory.

Logic in one view deals with means-consequences problems in which the means are procedures in inquiry and the consequences are warranted assertibility.[119] It follows that all phases of scientific method are materials of logic and that there is no separate log-

[118] Cf. Lindeman, *Social Discovery*, pp. 54, 56. Note that Webster's dictionary distinguishes *example* from *analogy* in that in the former "we argue from the mere similarity to two things; in the latter, from the similarity of their relations." *Webster's New International Dictionary* (2nd edition, Springfield, Mass., 1935), p. 94.

[119] Dewey, *Logic*, Chap. I, and pp. 432–41.

ical method apart from any other scientific method. To say that an assertion is arrived at logically is to say that it meets the current standards of operational inquiry for its subject matter.

In a narrower view, logic as classical logic comprises a method and standard of conceptual reasoning. In that case logical method refers to a part of a full inquiry. It refers to the handling of directing concepts. Such operations are a part and a continuing aspect of every inquiry; but they function as instrumental in devising tests and not as tests in themselves.

STATISTICAL METHOD

The statistical method has by all odds enjoyed the greatest favor among writers on social science method. There can be little doubt that social scientists have given greater effort to sharpening and refining the techniques of statistical analysis than to any other aspects of the full range of scientific method. Social scientists have tended to believe that statistical devices were their primary inheritance from the natural sciences. Yet the use of statistics has also been under heavy fire as has been pointed out previously.

No doubt a share of the conflict surrounding statistics is a conflict, not over the use of quantitative data but over the particular ways in which such data are handled. Thus, the 1928 committee of agricultural economists criticized the tendency to amass great quantities of data, yet itself put emphasis on statistics as a scientific method that is preferable to other methods of inquiry. In view of the often vague definitions of statistical method (such as that it is the use of mass phenomena quantitatively stated), conflicts can and do arise that do not concern the failure of the method thus defined, but that concern the devices used to handle such phenomena. For the purpose of this paper, it will be more fruitful, then, to discuss statistics in terms of its functional role in the full range of inquiry.

An important source of difficulty in respect to the debates over statistics arises from the professed aims of inquiry. If the objective of inquiry is, as Pearson stated, to form descriptive generalizations among observed particulars, then the primacy of any devices constructed to make summaries and to describe them briefly is obvious. If the objective of inquiry is to resolve problems in experience, such devices still fill an important role, but that role is not the end-in-view of inquiry; it is only instrumental in the full process of inquiry.

A clear testing block on this point is provided by the question of causal relations among particulars found to have a high statistical correlation. Every statistician of repute agrees that high correlations are not in any sense to be interpreted causally. But it is also clear that if research results are to be useful in experience, some causal understanding is necessary. The choice is then either to admit that statistical correlations are not intended to explain or to be useful, or to find some way to relate statistics operationally to other steps in inquiry. This dilemma is recognized in the 1928 handbook, not, however, because inquiry is seen as problem-solving but because the desirability of prediction is predicated. Nevertheless, the handbook recognizes the importance of showing that statistically determined "relationships are those of cause and effect as well as of covariation." But "Just how this can be accomplished, unfortunately, cannot be definitely stated. Here the economist will find in each individual case (i.e., research inquiry) an individual problem which will tax his powers to the utmost resort to every means available to him to discover what may be the causal relations involved. To a large extent it may be necessary to fall back upon the logic of *a priori* analysis."[120]

This concept of scientific procedure fails to comprehend that logic as conceptual reasoning and statistics as correlation of abstracted static particulars fulfill the same office: that of suggesting further search within the experiment, or unit of action, to close the gaps in interaction and sequence within the observed experience.

This discussion then leads to a consideration of the role of statistical devices to reveal evidence of relationship of particulars as such or evidence of relations within experience. And this consideration introduces a distinction between the technique of classical statistics and control statistics.

As a result of efforts to use statistical devices in industry "to establish ways and means of making better and better use of past experience,"[121] a relatively new area of statistics has been developing particularly in the past 15 years. This movement has achieved its greatest advance in respect to controlling manufacturing processes. It has begun to affect the use of statistical devices in respect to controlling experimental processes in physical science. It has barely begun to affect the field of social science.

The fundamental differences between classical statistics and control statistics are so great that they have to be considered in different categories. Classical statistics deals with measurements drawn at random from a sample that, for the purpose of the statistical devices used, is static and isolated. Control statistics deals with sequential

[120] *Research Method and Procedure*, pp. 275, 281.

[121] W. A. Shewhart, *Economic Control of Quality of Manufactured Product* (New York, 1931), p. 351.

relations that are not drawn blindly from a static sample but are kept under constant observation within units of action. In other words, the behavior attributes of a unit of action are kept intact as to their interaction and process relations in sequential or control statistical devices.[122]

To the extent that use of the new techniques of control statistics closes the gaps between observed data within units of action, it provides a testing force. Statistical method in this new form tends to resolve the constant confusion in respect to correlation relationships and causal relations in data. If *causal* has a useful meaning, it is that action undertaken causes the intended consequence; and such relations exist in the experience of an acting unit. They may be suggested by establishment of conjunctive relationships in classical statistics, but they are found in experiments in the form of sequential relations, which control statistics insists must be preserved.

PROCEDURES FOR PROCESSING EVIDENCE

In discussion, methods of research are often confused with techniques or procedures of research. There also has been some debate as to the superiority of survey, participant observer, library, accounting, cost route, interviewing, case history, mail questionnaire, and other techniques for obtaining evidence. The extent to which any of these techniques will result in useful evidence depends not only upon the skill that is used in recording and handling the materials secured but even more upon the nature of the information and its relevance to a formulated problem and a hypothesis.

By any technique, some data can be obtained that can be used for finding the presence or absence of abstracted relationships among particulars. For this phase of inquiry the important point will be the adequacy of the statement of the hypothesis so that important gross checks of affirmation and negation may be made immediately.

The securing of materials that have testing force is not as easy. It is clear that the more nearly the scientist is a direct observer of the

[122] Abraham Wald points out that "sequential sampling" is "essentially different" from "single sampling" as he calls it in "Sequential Method of Sampling for Deciding between Two Courses of Action," *Journal of the American Statistical Association*, vol. 40, 231 (Sept. 1945). W. Edwards Deming differentiates Type A and Type B classifications of statistical inference problems and notes the "*Necessity for keeping in mind the ultimate objectives of an inquiry*" whether interest "centers in the product" (Type A) or "*in the process* . . . that gives rise to yesterday's, to-day's, and tomorrow's product." In the latter case, Type B, "a large batch of data is studied, not simply as a large sample, but as an ordered sequence of small samples." "On A Classification of the Problems of Statistical Inference," *Journal of the American Statistical Association*, vol. 37 (June 1942). Cf. W. A. Shewhart, "Contribution of Statistics to the Science of Engineering," in *Fluid Mechanics and Statistical Methods in Engineering* (University of Pennsylvania, Philadelphia, 1941).

experiences he is using as experiments, the more likely he will be able to uncover closely-woven sequences of strategic elements in each of his experiments so called. It is because of this possibility and its importance in testing in social science that the participant observer technique has strong advantages.

Whether case history, survey, or interview techniques produce evidence with testing force depends upon whether the data are obtained as a part of a study of a clearly formulated problem, whether the hypothesis is used at the moment of recording information, and whether the aim is to record an interacting and sequential set of data or whether it is to record a bill of more or less isolated particulars.

SUMMARY

If research is seen as a process of inquiry, with problems emanating from, and tested in, experience, with generalizations serving instrumentally to suggest possible causative connections in experience, and with tests dependent upon such relations in the evidence of experience, then many confusions in the traditional views of research methods tend to clear up. The conflicting claims for each procedure find a place in the outline of a full inquiry; the methods themselves fade away. What results is a full concept of modern science; and the techniques or devices previously held as constituting a scientific method turn out each to have a function to perform in the one task of processing evidence in order to arrive at a proposed method of resolving an existing problem in experience.

In this view, an ideal of absolute perfection and absolute certainty gives way to that of intelligent judgment progressively using the results of experience to suggest actions by which intentions and results are as closely united as man can make them in an ever changing world.

CHAPTER IV

On the Use of Publications as Methodological Evidence

Before entering upon analyses of various published results of research, which will comprise the following chapters, certain introductory remarks are in order. These comments are necessary, first because specific criticism of published studies is uncommon in rural social science, and critical reviews are therefore easily subject to misinterpretation. Secondly, the use of published reports as evidential tests of research method has certain limitations that need to be recognized.

It is important to make clear that there is no intent or need in the following chapters even to imply an evaluation of the abilities of the authors of the studies used. In the first place, in view of the conflict and confusion that have been shown to exist in the entire arena of social research method, it is hardly conceivable that any individual could find indisputable guidance as to research concepts and procedures. Secondly, since it is impossible and unnecessary here to make equally intensive analyses of all publications in rural land economics, it would be erroneous to attribute any special connotation to the particular authors whose work is scrutinized in these pages.

Similarly, there is no intent or need to impute an evaluation to the publications as publications. There may be a number of reasons other than the promotion, or distribution of the results, of scientific research to justify any given publication. These publications are used only insofar as they provide experimental evidence for a study of research method; there is no attempt or interest here to discuss aspects or purposes unrelated to this purpose.

The foregoing remarks should depersonalize the subsequent chapters. But they also raise considerations in respect to the limitations on the use of such experimental evidence as is available for a study of this kind.

One limitation is that hardly any publication of research results presents a detailed chronology of procedures followed. The present study is itself illustrative, for, as it was stated in Chapter I, this study is not organized around the chronology of author's work on the problem over the period of a decade. If the cross-checking and backtracking were spelled out, it would discourage readers without adding to their enlightenment. Since publication is arduous to the author and expensive to the publisher, and since a reader's patience is hardly boundless, research publications are usually shaped to give an efficient, concise report rather than a complete itinerary of the investigator's wanderings.

These factors prevent publications from being complete evidence for our purposes, but they do not invalidate such evidence — and they may strengthen it. A publication may be streamlined, yet it generally represents an organic unity, a coherent exposition of the scientist's efforts in inquiry. Such items as may be omitted are what the author judges (within certain other limitations discussed below) can be deleted without destroying the internal structure of his work. On this basis, the trimming that is done should result in the presentation of the clearest problem formulation, the most conclusive arrangement of evidence, and the most significant conclusions that are made possible as a result of the research done.

Yet there are other limitations. One of these is that a published study may not be intended to represent anywhere near a full inquiry into any given problem. Since a complete inquiry is often an extremely large task, there almost always is need sometime to decide at what stage the then-attained results should be published. There may easily be substantial reasons for printing a report even though the study is not definitive. But these possible limitations on any particular publication are not damaging for the purposes of the present investigation. An admittedly partial study still can provide evidence on methodology. The analysis of course must recognize the limited setting of the research endeavor in question. In turn, however, that limited setting itself may be analyzed in terms of its orientation toward the larger questions to which it is supposed to be relevant.

Another reason why published reports do not always represent full inquiries is that there always are some administrative, temporal, spatial, or financial limitations on research projects. Again, however, these restrictions do not destroy the value of the reports as evidence of research procedure, but they do add further reason for emphasizing that the purposes of this review do not involve a general evaluation of the publications referred to as publications. If the purpose of a critical review of a research publication were merely to criticize such reports on the basis of their failure to attain a perfectionist ideal, many of them would be defended on the grounds of these practical limitations. But for the purpose of finding out what procedures lead to what results, there is something to be learned from all research work, regardless of the practical limitations on any particular study, and no need to take either offensive or defensive attitudes toward such analysis. Especially since every future research study will also operate under some practical limitations, it would be unwise to disregard the lessons that can be learned from past work, merely because it was conducted under circumstances that may have been more limiting than the investigators would have wished.

Another point that must be borne in mind is that all publications, even though sponsored by agencies receiving funds for research, are not meant to present the results of scientific analysis in any formal sense. Nor should disapprobation be attached to a publication merely because it is not the outcome of scientific inquiry. It is important to the training of scientists and to the advancement of the processes of inquiry to distinguish between publications that report scientific investigations and those which do not. But these distinctions should be drawn without any implication of status, one as to another. There is an infinite variety of activities which are not scientific inquiry, but they do not automatically stand in need of justification on that account. If, for example, some persons like to draw charts to satisfy some immediate taste in the same way that another may like to write a certain type of poetry, it is no condemnation if it be said that such making of charts is not a part of scientific research. But it is important, for the purposes of a study of scientific method, to recognize whether or not that making of charts is taken to be a part of a problem-solving inquiry.

A further possible limitation on publications as methodological evidence is that imposed upon research workers in social science. They study the behavior of people who may sometimes prefer not to have their activities laid open to public gaze or who may be impeded in their pursuit of self-interest by the recommendations proposed as a result of scientific investigation. Consequently, problems may not be formulated as clearly as they might be, all the evidence known to the student may not be inserted in the printed report, and conclusions may be stated in a backhanded manner.

This limitation would be severe if research workers were evaluated as scientists by an analysis of their published work. It can be argued that they know more than a particular publication indicates. But, again, such is not the purpose of the present inquiry. The limitation, however, still must be recognized as detracting from the usefulness of the evidential materials available. Yet it should also be understood that if these considerations so modify the publication of research results that the reports cannot be safely used as scientific reports, then it would also be true that they could not safely be relied upon to promote public understanding and action upon social problems. If this is the case, it would constitute a professional and a public service to indicate the extent of the deficiency.

Finally, it should be made perfectly clear that the following analysis of specific research reports is made as only a part of an attack on a problem, and is undertaken, insofar as possible, with a scientific attitude, which, after all, is the mark of the scientist as the experimental method is the mark of modern science. The writer has engaged in enough research projects in the past twelve years so as not to have been excepted in his own work from the confusions that have somehow borne upon every research worker. It happens that during this same period, he has made a special effort to look into certain of the problems that plagued him along with many of his colleagues. If the present effort has a positive contribution to make, it will be one that the writer has felt the need of and has been seeking as much as anyone else.

Selection of Evidence

A study of methodological problems in research itself involves decisions in respect to the processing of evidence, which decisions in sum determine the methodology of the inquiry.

As indicated previously, the evidence for this research is the published results of accomplished investigations in rural land economics. The complete literature of such reports does not of course comprise the entire range of possible evidence. The full range would include accounts of abortive research — that which was started but failed to see the light of day either because it was recognized as inadequate, or being thought adequate for publication, was not published for any number of reasons. It also includes journal articles and other accounts of research efforts and methodological discussions.

Of these various sources, the published results of research and the methodological accounts and discussions have been combed for the present study. It would not be possible to affirm that every last bit of such evidence has been analyzed, because it is always possible that some items would slip through the comb. It can be affirmed, however, that a diligent effort has been made to scan all the material available.

Unpublished research is even less available as evidence. Even if the investigator is familiar with such work, it is practically impossible to refer others to it. The only claim for having made use of such evidence in this study is that given in the first chapter. The writer has had an opportunity to be close to scores of research efforts in their infancy and growth, many of which never resulted in a printed publication. Because of his continuing special interest in methodology, he has, over the years, used these experiences with abortive research efforts as testing grounds for the then-formulated hypothesis concerning research problems in rural land economics. It would be inaccurate to say that such research has not entered into the work that lies back of this presentation; but it is true that such evidence is not here presented, and that only a small part of it could possibly be available to this or any other one investigator.

This report, then, will offer as evidential tests, published results of research. The particular publications used in this presentation will be selected from among the whole literature that has been reviewed. Since the nature of this type of experimental evidence involves in its presentation a great deal of space, it would be more than a monumental effort to cover it all in writing. What is more, it would be unnecessary insofar as it would in many instances be very repetitive.

In view of these considerations, an effort will be made to present analyses of research that covers a wide range of land economics problems, that spans the time during which land economics has grown to its present status in the United States, and that has been done by a variety of research personnel with varying backgrounds of training and experience. This diversity as to subject matter, time, and schools of thought should prevent the possibility of bias and show, both positively and negatively, the impact on research of the methodological concepts that have been previously discussed.

It is further intended in the following chapters to refer to numerous publications to which the point being made, with reference to any given publication, is applicable. Furthermore, while some studies will be discussed in considerable detail, related studies will be more briefly discussed in the same section. In these ways, the reader may see that the present formulation of the problem is warranted and that the experimental evidence in respect to the present form of the hypothesis is substantial.

CHAPTER V

Land Utilization Research in the Cutovers

It was brought out in Chapter II that the study of problems associated with major changes in the pattern of economic use of land has been important in the development of rural land economics. Land utilization research can be described as dealing with problem situations in which people in a given locality are in the process of transformation from activities with certain land requirements to activities with different land requirements. As examples situations may be cited in which arid lands are being put under irrigated cultivation, logged-off timber lands are being settled, farms are being abandoned, or farm land is being transformed for urban types of uses.

Before 1920, there were two general situations in which land previously unused for farming was being settled for cultivation: the arid lands and cutover timber lands. It was in the arid regions that the need for special public assistance to settlement was first felt, as evidenced by the Desert Land Act, the Carey Act, and particularly the Reclamation Act of 1902 with its subsequent amendments. Irrigation projects, however, were primarily planned and directed by engineers. Although Professors Ely and Taylor both spent some time on irrigation economics problems, no literature on irrigation economics research was developed until after World War I.

The settlement of arid lands under dry farming occurred in the same period and involved some special federal legislation like the Kinkaid and Enlarged Homestead acts, but the economists did not study these problems to any extent.

In respect to the settlement of cutover lands, however, men in the developing fields of farm management and agricultural economics undertook specific research projects. Out of these efforts has come a long series of research studies, publications, and public action. To the inquiries conducted in these problems, first attention will be given. Furthermore, because of the high degree of attention to these problems in the northern Lakes States and because of the close connection between the history of land economics and the land-grant institutions in this region, a much more thorough review of the reports on the

Lakes States cutover region will be given than for reports on other areas or other types of land economics research.

The Lakes States Cutovers

Under the agricultural conditions which prevailed between 1900 and 1920, there was continuing pressure to settle unused lands for agricultural purposes. In the upper Lakes States, the white pine had been removed in the latter decades of the nineteenth century; then the lumbermen went over the area once again to take the hardwoods.

At first, the expectation was that most of this land would be turned into profitable farms. But as early as 1915 it was seen that in spite of generally favorable conditions for farmers in the country as a whole, some settlers in the cutovers were having real difficulty in establishing profitable farms. The most obvious special difficulty was that of removing the stumps from this land; and just as engineers in the respective land-grant colleges worked on land-clearing techniques, so did some farm management workers study the costs of clearing operations. The problem was seen as that of helping the individual settler to determine ways and means of clearing his land inexpensively and quickly.[1] In some instances, average cost figures were calculated from schedules taken from settlers using different clearing methods; in others, by presenting a series of case reports on the costs incurred on farms with different clearing obstacles or using different clearing techniques.

U.S.D.A. BULLETIN 425, 1916[2]

In addition to the work of calculating the costs of land clearing, some general farm management surveys were begun. In 1914 the U.S. Department of Agriculture arranged to take about 800 schedules summarizing the farm business situation of settlers interviewed in three states in order "to discover the more profitable farm practices and the factors essential in the development of the region."

[1] On the front cover of an early bulletin it is stated: "There is needed only the intelligent effort of earnest and ambitious settlers to convert thousands of undeveloped tracts in Upper Wisconsin into profit-producing farms." E. J. Delwiche, *First Aid to the Settler* (Wisconsin Agr. Expt. Sta. Bull. 260, 1915). Also see A. J. McGuire, *Land Clearing* (Minnesota Agr. Expt. Sta. Bull. 134, 1913); H. Thompson and E. D. Strait, *Cost and Methods of Clearing Land in the Lakes States* (U.S.D.A. Bull. 91, 1914); M. J. Thompson, *Investigations in Cost and Methods of Clearing Land* (Minnesota Agr. Expt. Sta. Bull. 163, 1916); B. G. Packer and E. J. Delwiche, *Farm Making in Upper Wisconsin* (Wisconsin Agr. Expt. Sta. Bull. 290, 1918); M. J. Thompson, *Forced Against Delayed System of Clearing Stump Land* (Minnesota Agr. Expt. Sta. Bull. 189, 1920); John Sweenhart, *Clear More Land* (Wisconsin Agr. Expt. Sta. Bull. 320, 1920); H. L. Russell, *Farms Follow Stumps* (Wisconsin Agr. Expt. Sta. Bull. 332, 1921).

[2] J. C. MacDowell and W. B. Walker, *Farming on the Cut-Over Lands of Michigan, Wisconsin, and Minnesota* (U.S.D.A. Bull. 425, 1916).

The bulletin is a very good example of a standard pattern of research that grew out of the procedure for farm management survey. The analysis consists of averages and some frequency distributions of various items of the farm business for all 800 farms lumped together. Beyond this, the farms are grouped into tillable acreage classes and the average incomes of the classes are compared. For a selected group of dairy farms, comparisons are made of the average incomes of farms classed by tillable acres and subclassified by income per cow. Then, all the farms are classed as those above or below the average crop yield and their average farm incomes are compared. In each case, average incomes are higher for the groups with the higher tillable acreage, crop yields, or income per cow.

If the stated purpose of the study were set in terms of a beginning hypothesis is more than a tautology. Within the framework of a cessful farmers they would have more profitable farms. On the basis of what was done, the "if" clause may be said to have been refined to suggest that if unsuccessful farmers increase their tillable acres, crop yields, and income per cow, then they would have profitable farms. The results appear to substantiate this form of the hypothesis as a conclusion.

In the first place, however, there is question as to whether this hypothesis is more than a tautology. Within the framework of a pioneer economy, even a modest tillable acreage may itself be an indication of successful farm settlement since a lower acreage simply is no farm at all.

In the second place, the postulated factors making for success are not direct actions to be undertaken but are themselves the results of direct actions. A farmer cannot just decide to have more income per cow. The best that could be hoped, therefore, from this type of analysis is a suggestion for the next step in an analysis to solve the problems. In the report, this factor is taken to mean that the farmers should have higher-grade cows, but this is a casual suggestion only; it is not based on evidence that higher-grade cows are strategic to higher income per cow. Another illustration of this methodological limitation is in respect to increasing tillable acreages. The number one conclusion of the inquiry is this: How to enlarge the tillable area economically is the first and most important agricultural problem of this region. This conclusion has scientific status, precisely because it recognizes that the analysis of tillable acreage is only suggestive of a new hypothesis and is not conclusive with respect to the initiation of action. In other words, the inquiry does not answer the question *why* the profits of some farms are low by showing *how* to act to increase them.

In the third place, the data are not arranged as evidence of the direct experience (experiment) of the settlers of the area studied. Had the experience of each settler of the process of settlement been preserved as an experiment, the conclusions might have been different. Also, they might have been more conclusive with respect to action to be taken. For example, it is known that some settlers were first- or second-year settlers. If these settlers were the only settlers with small tillable acreages and were the only ones with low farm incomes, and if the longer-time settlers were also those with larger acres and also the only ones with high farm incomes, then the time element itself would be the solution. Again, it is known that some settlers had unsuitable soils. Were it shown that these settlers, regardless of length of settlement and regardless of acreage, all had low incomes, then the idea of increasing yields or income per cow would be quickly discarded as a futile one. Finally, it should be noted that under this form of analysis, the exceptional cases would require explanations, and this procedure would serve to refine and modify the hypothesis.

In the fourth place, the scientific status of this inquiry as only a very preliminary probing of the problem situation is further indicated by the general suggestions offered as a part of the outcome of the study. These suggestions are, in part, that the settlers should select farms with good soils, that they should get locations near market outlets, that they should avoid paying too much for their land, and so forth. None of these factors is used anywhere in the analysis of the evidence. They are clearly based on the casual observations of the authors, yet they are the only direct suggestions offered relative to the institution of new actions in the processes by which individuals try to establish farms on cutover land.

MINNESOTA BULLETIN 180, 1918[3]

Another farm management study of settlers on cutover land was published by the University of Minnesota in 1918 under the title, *Experiences of Northern Minnesota Settlers.*

That this study also deals with a problematic situation — with doubts as to the outcome of a process of experience over time — is made clear in the report. The situation dealt with is such that the "average settler" is "usually unable . . . to bring a living for his family and to permit him permanently to stay in the region. Many

[3] F. W. Peck, *Experiences of Northern Minnesota Settlers* (Minnesota Agr. Expt. Sta. Bull. 180, 1918).

farms have changed hands two, three, and even four times with only 25 to 30 per cent of the possible agricultural land broken for crop production." Thus, the action taken with a view to establishing permanent and prosperous farms had been far from harmonized with the outcome.

There is no further refinement of this problem, nor is there any statement of a beginning hypothesis. The only specific statement on the objective of the study is in a paragraph which reports that prospective settlers are seeking to learn of the chances of "making a home and a respectable living." "In short, they want to know the experience of others, for that is the only source of reliable information."

In pursuit of this objective, forest rangers in ten districts collected information regarding the experiences of more than 200 settlers. Of these schedules, 141 were found usable and the bulletin presents the data from them. Information was obtained concerning the date, the cash on hand, and the acreage cleared at time of settlement; the acreage cleared per year since settlement and the estimated cost per acre; and the total acreage, cleared acreage, inventory of livestock, kind of crops grown, and sources of income at the time of survey. Also, each informant stated what he felt to be the dominant settlement problems and what ought to be done about them.

The data collected in the survey are presented by topics; and for nearly all topics there are tables showing the number, average, range, or percentage for each item by districts and for the entire group of 141 schedules. These tables are accompanied by a written summarization. In Part II of the report there are other tables, without written summarization, which give the data from each schedule in each district so that "All the facts for any one farm can be assembled by following the farms by number through the set of four tables for each county group."

With these presentations of summarized district data and individual farm data, the bulletin certainly makes available, in line with its stated objective, information regarding certain aspects of the experience of the settlers in the cutover region. Looked at in terms of solving a problem, however, what about the study as scientific inquiry? That there is an intent to help solve the problematic process of settling cutovers is explicitly shown in a section entitled "Conclusions." There is no question but that these conclusions are actions which are proposed to be instituted in the settlement process so as to secure more profitable and permanent farms, the end-in-view of the settlement process.

The conclusions of the bulletin may be formally restated in four categories and in "if-then" form: If (1) land utilization and soil surveys were made, if publicity were given to ownership of good land, and if real estate dealers were licensed to assure settlers' applying their efforts to good land; and if (2) public assistance in clearing and draining lands were given to help prepare the land; and if (3) settlers would recognize the limiting nature of markets, develop livestock enterprises, manage their woodlots, and avoid work off their farms; and if (4) research were conducted on clearing of land, marketing, type of farming, and colonization problems—then more prosperous and permanent farms would be created in northern Minnesota.

In view of the problematic nature of the situation dealt with and the problem-solving nature of these conclusions, what is their scientific status on the basis of the procedures revealed in the report? An analysis of the evidence given in the publication indicates that these conclusions are not warranted so far as the research in the bulletin is concerned. Their only status as scientific inquiry is that they pose a formulation of the problem and hypothesis yet to be tested by experimental inquiry.

Let it be supposed that a beginning formulation of the hypothesis had suggested that lack of information respecting good sites was a strategic factor in failure and that, conversely, information pointing to good sites would assist in achieving success. Then, if this hypothesis were used to direct the processing of evidence, clearly there would have to be some attempt at least to show that settlers on poor sites failed and settlers on good sites (with or without the intervention of other factors held to be possibly strategic) succeeded. In the report, however, there is no reference nor any data anywhere on site quality or on the process of site selection. Without such data there could not be any sorting of the settlers by their actual experience in respect to this question, even were it intended. In other words, without a working hypothesis there was no indication that site quality facts should have been sought out as evidence.

But even where there are some data relevant to the conclusions, they are not handled in a manner that reveals their relevance to the problems of the settlement process. Routinely, the evidence is treated on the basis of geographical districts; and there is no indication of significant differences among the districts as to the presence, absence, or variation of factors considered strategic to successful settlement.[4] Hence, there is no chance for testing by comparisons and contrasts

[4] The only difference noted is that one of the ten districts is said to have more prairie land than the other nine.

among the settlers' actual experiences (experiments) in respect to possible means for redirecting settlement to desired consequences. Such an analysis, for the items included in the schedule, could be partly garnered by checking the possibilities with actual settlers' experience, insofar as it exists in the individual farm data given in Part II of the bulletin. But no such analysis is given as a part of the research presented.

One further illustration of these methodological limitations should be sufficient. In respect to off-farm work, it is stated that "During the last few years, outside labor has become relatively scarce. This has undoubtedly had a good effect upon land settlement, for the men have been forced to clear the land and work it into a more productive condition." Also it is stated that "Receipts for labor performed away from the farm decreased in amount and importance for the better class of settlers. State aid in land clearing would eliminate largely the necessity of this kind of work." Evidence necessary to warrant these statements as conclusions would, at a minimum, require facts to be arranged to show, first, that off-farm work had decreased over time; secondly, that on those same farms where the decrease had been felt the surplus labor time had been devoted to land clearing; and third, that the resulting increase in productivity on those same farms was greater than the loss in off-farm income. In contrast, the data given show only the amount and percentage of outside labor income for one year for each farm and an average for all farms in each district.

The level of these figures may suggest something about outside labor receipts in the settlement process, but it cannot serve as evidence of the sequential interpretation that is given. To the extent that the opinions expressed by settlers may have substantiated the interpretation, a measure of weight is added to it. But for our purposes, the point is that the significance of the interpretation as a conclusion rests upon such opinions and not upon the outcome of the empirical analysis.

In summary, this bulletin presents both a problematic situation and conclusions the characteristics of which are consonant with those given in Chapter III. Between these phases of inquiry, however, the collection and processing of evidence are out of gear. The absence of data absolutely necessary to substantiate certain aspects of the conclusions and the lack of arrangement in the available evidence so that it is relevant to the conclusions indicate that so far as a full inquiry is concerned (1) there is no development of a beginning hypothesis to steer the course of processing the facts as evidence (a tendency merely to summarize and generalize data in a formal way), and (2) the scientific status of the conclusions is that they represent a formulation of the problem and of a beginning hypothesis.

WISCONSIN BULLETIN 318, 1920[6]

The first research publication dealing with the problematic process of cutover settlement prepared by men whose major training was in general economics, as distinguished from those who specialized in farm management, was published in Wisconsin two years after the Minnesota bulletin appeared. By this time, the problem apparently was clearer, for the Wisconsin study is directed at one of the particular factors then considered strategic in the situation: the availability of credit for settlers. Farmers, county agents, bankers, and others reported that lack of capital and credit is a "chief cause of failure." The study is obviously intended to revise and expand a hypothesis that if changes were made in the farm credit arrangements in the cutover area, settlers would be better able to establish profitable farms.

In strong contrast to Minnesota Bulletin 180, the organization, analysis, and presentation of the evidence in this bulletin would very properly be described as informal. There is no attempt whatever to use in any consistent manner any particular sets of evidence; there is not a single table of data in the report or any indication that such tables were assembled; and the only map, one which shows interest rates by counties, is based on "reports of several bankers in each county." In short, the bulletin is an exemplar of the so-called "Wisconsin essay" research report, as Minnesota Bulletin 180 is of the farm management survey.

There is no formal statement of procedures used, but the following quotations, taken from various places in the bulletin, tell the story: "The testimony of the county agent, the bankers, the real estate dealers, and the successful farmers agrees that . . ." "In answer to a questionnaire, a majority of the land companies said that . . ." "The facts in the case, as brought out by almost unanimous opinion of the county agents . . . and others, are that . . ." "The store accounts of between 300 and 400 settlers plus the opinions of a great many merchants and farmers show that . . ." "The settlers talked with and those who answered a questionnaire relative to the matter almost always said that . . ." "In order to have as true a picture as possible, first-hand observations were made in various sections. These were supplemented by the testimony of . . ." "It is preeminently the opinion of the experienced farmers and others who have made a study

[6] R. T. Ely, B. H. Hibbard, and A. B. Cox, *Credit Needs of Settlers in Upper Wisconsin* (Wisconsin Agr. Expt. Sta. Bull. 318, 1920).

of conditions . . . that . . ." "This statement was also verified by house-to-house canvass of what was taken to be a typical community of about 25 such settlers."

Not only are the sources of evidence vague, but the presentation even of quantitative materials is noticeably inexact. Thus: "a fairly good house, barn and other buildings would have cost before the recent sharp advances in prices not less than $3,000; fences not far from $300 . . . Thus . . . he will need to have . . . some $4,000 or $5,000 in his investment, plus whatever is required over prices and costs of 1917 or 1918."

How are these informal materials, collected from a hodgepodge of sources and in a variety of ways, used as evidence? They are organized around an analysis which, when outlined, presents an orderly set of relevant concepts. The first of three parts of the bulletin explores the "credit needs in the pioneering stage" for buying raw land, getting equipment, clearing land, and for meeting living, operating, investment, and other expenditures. The second, entitled "Financial Needs for Farm Development," deals with the difficulties involved in getting from the pioneering to the developmental stages of settlement and includes sections on the consequences of inadequate credit and the investments involved in a developed farm unit. The third part reviews the existing credit agencies, the types of credit available to settlers with money on land, and then deals at greater length with the problems of settlers "with little money."

This bulletin has no particular section labeled conclusions. Here and there throughout the report, suggestive comments are made. In terms of the objectives of the study, however, the analysis, as indicated above, works its way down to segregate the source of the problem in the group of settlers with inadequate means. But this sorting is a classification of conceptual groups, not of any given set of reported experiences. The report indicates why the usual credit channels do not meet the needs of this type of settler and suggests three things: If outside capital (for example, Federal Farm Loan Banks) can take over real estate loans, then local capital will be available for personal credit; if a plan like the Ashland Dairy Plan is used, then more money will become available locally; and if the progressive land companies get the good will and encouragement of the state, then the settler can get useful services along with needed credit.

On the basis of the evidence presented in the bulletin, what is the scientific status of these conclusions? The answer to this question must be that such conclusions as are offered in this bulletin are not grounded by direct empirical evidence. They are based, so far as the report shows, solely (1) upon the conceptual reasoning in the text, (2) upon statements unsupported by any factual materials, or (3) upon statements supported by materials that may be factual but that are not subjected to any experimental test.

That the entrance of the Federal Land Bank funds will make for easier personal credit is expressed only as a "most hopeful outlook." It is reported that these banks have put 17 million dollars into the pioneer districts, but there is no evidence as to the consequences that have been effected in such areas in respect to personal credit. Similarly, with regard to experience with the recommended Ashland credit plan, there is only a solitary statement that in "a few communities" it is making "splendid progress." Nor is there anywhere in the report any evidence that compares the experiences of settlers with and without loans from land companies, be they progressive or otherwise.

The methodological inadequacies of this study are not merely in the informality of the collection and handling of evidence. When the bulletin states that by talking with informed people and by going over some mailed questionnaires it is judged that something "in the neighborhood of" so many dollars is needed for equipment, the figures probably serve as good an office as though averages were calculated to the last decimal point for a large number of farms. The difficulty is rather that while the orientation of the report around a conceptual analysis of the role of credit in the settlement process helps to clarify the problem, the scantiness of empirical evidence makes the burden of conviction to rest almost entirely on the conceptual reasoning. Furthermore, because of the informality of the evidence, which is introduced from different sources in a seemingly haphazard manner, the factual references do not serve to test the concepts. Since the data are only used where helpful for purposes of exposition, there is no assurance that experiential materials would support only the argument of the report.

In summary, this bulletin presents both a problematic situation and conclusions the nature of which are consonant with those given in Chapter III. Between these phases of the inquiry, there is a presentation of materials that are treated in terms of the elaboration and refinement of a beginning hypothesis that is germane to the problem as formulated. The conclusions, however, cannot be regarded as warranted assertions, not because the materials offered are irrelevant to the development of the hypothesis, but because they are not consistent as evidence. Ranges of experience are not given to provide tests of the conceptual reasoning. The complete absence of data absolutely necessary to substantiate certain aspects of the conclusions and the lack of arrangement in the available evidence so that it is relevant to the conclusions indicate that so far as a full inquiry is concerned, (1) there is no development of consistent experiential evidence in the course of developing the hypothesis (i.e., a tendency merely to reason conceptually), and (2) the conclusions can only serve as a tentative formulation of the problem and a beginning hypothesis for future inquiry.

MINNESOTA BULLETIN 196, 1921 [6]

In 1919 data were collected by the Minnesota farm management division from farms in two Minnesota cutover communities and one prairie area "for the purpose of comparing the rapidity of farm development in the prairie and the cutover lands" and "with a view to learning the conditions most favorable to the development of farms in the cutover region." While the inquiry is thus directed toward the same problematic situation as the preceding ones, there is no further specifically formulated statement of the problem, or working hypothesis.

The bulk of the report consists of an item-by-item summary for each area separately of the farm business data secured in the survey, and the written report in general states these figures as an average for all farms in each area and explains how they were calculated. In a final section, these average business items for the three areas are set up side by side and some of the differences in the averages are noted.

An indication that a beginning hypothesis was entertained during the study lies in the fact that all the farm business data for each county are averaged on a cross-classification by acreage — total farm acres in the case of the prairie area and acreage cleared in the case of the cutover counties. Also the final pages emphasize that the rate of land clearing is "a matter of failure or success" to the settler. In the written analysis, however, there are hardly any references at all to the lengthy tables in which the cutover farm data are averaged by amount of land cleared, and were these tables used, they would cast doubt on the suggestion that acreage cleared is a crucial factor in the settlers' success.[7] From this, it appears that a beginning hypothesis

[6] C. G. Worsham and A. Boss, *Farm Development Studies in Northern Minnesota* (Minnesota Agr. Expt. Sta. Bull. 196, 1921).

[7] The group averages presented in the bulletin show that in Beltrami County the group with the smallest cleared acreages had an average labor income higher than the next two groups and almost as high as the third group above; and in Itasca County the group with the smallest cleared acreage had a higher average labor income than the next four

was used to set a pattern of cross-classified averages, but that when this pattern of data did not support the hypothesis, instead of working through a modification of it, emphasis upon it was omitted from the written analysis but retained in the conclusions.

Another general conclusion of the report is found in a paragraph that states that "the settlers are laying the foundation for substantial homes and businesses. In other words, the settlers are encountering and solving perplexing questions and situations as they arise." This statement is difficult to understand — except for the fact that the data were collected in a highly prosperous farm year — in view of the other studies of cutover area problems, the introductory remarks to the report, and the three paragraphs in the bulletin in which are listed the problems of settlement as reported by the settlers.

One other statement that may be interpreted as a conclusion is the emphasis in the introduction on the dangers of settling on poor rather than good cutover land. However, there is no direct evidence that applies to this suggestion in the entire report, unless it is that the description of the Itasca area emphasizes poor quality land whereas that of Beltrami emphasizes the good qualities of the land. In the section in which the area data are compared, however, the text only compares the cutovers with the prairie area, but the averages for labor income, farm income, and total farm returns are higher for the Itasca area than for the Beltrami area.

In sum, this bulletin illustrates the analysis of data in such a way that it not only fails to support its own conclusion but does not even allow one to see how it bears on other studies also dealing with the same large problematic situation. To begin with, it is not even clear how the problem was formulated or what the working hypothesis might have been. What seems to have been a beginning hypothesis was used to establish a set of tables of averages by which the data were generalized, but practically no use is made of these tabulations in the report on the study. Consequently, there are practically no conclusions in the report, and those which are made hardly seem even to achieve suggestive force so far as the evidential materials are concerned. All that can be said from the standpoint of scientific research is that data were collected and were summarized or generalized. No warranted assertions in respect to what to do to get what results are arrived at, and no suggestions as to needed modifications of a beginning hypothesis are found.

groups above, had higher average livestock receipts than the next group above, and higher average livestock products receipts than the next three groups above.

U.S.D.A. BULLETIN 1295, 1925 [8]

Cooperative research work undertaken in the cutover areas of the three Lakes States — Michigan, Wisconsin, and Minnesota — in 1918–1920 resulted in a research publication in 1925 in which the activities of land colonization companies received special attention as a strategic factor in the land settlement process.

The problematic situation — failure to establish prosperous and permanent farms on cutover land — is the same as that dealt with in the three preceding bulletins. The specific formulation of the problem for this inquiry follows the conclusion of Minnesota Bulletin 180 to the effect that agencies selling land should be regulated and land colonization activities should be further studied. In the opening sentence of U.S.D.A. Bulletin 1295, it is stated that "a settler's success . . . under present conditions . . . depends . . . upon the policies and practices of the land companies and other real estate agencies" which are "more likely to be animated by profit-seeking motives than by consideration of the public interest."

While the objective is specifically given, "to analyze the methods of the different types of land companies and other related agencies," it is clear that such analysis is intended as a way of making explicit a hypothesis that if the activities of the companies which intervene in the land settlement process were regulated, then settlers would achieve success.

To probe this hypothesis, data were secured relative to the experience of 3000 settlers and 153 land companies, real estate dealers, and agents. In addition, other items of evidence were obtained from secondary sources such as the census.

The conclusions of the inquiry are given in the final section entitled "Land Settlement from Standpoint of Public Interest." They may be stated as follows: If state or national machinery effectively eliminated purely parasitical types of agencies, if some method were employed to prevent the sale of land not suitable for settlement and to prohibit the sale of good land at "unduly high prices," if competition could be reduced so land companies would not make overattractive but costly inducements to settlers without adequate experience and resources, and if lumbering can be coordinated with settlement, then settlers will establish themselves successfully.

On the basis of the evidence presented in this report, what is the scientific status of these conclusions? The general answer to this question is that while these conclusions are, on the basis of the evidence

[8] J. D. Black and L. C. Gray, *Land Settlement and Colonization in the Great Lakes States* (U.S.D.A. Bull. 1295, 1925).

given, much more fully substantiated than those of the two reports previously reviewed, still the evidence is not as conclusive as might have been the case had the procedures of research been slightly different.

The central body of the report, covering over half of the bulletin, is based on a detailed analysis of 15 land colonization projects and the experiences of 583 of their settlers, selected from among the 153 land agencies and 3000 settlers from whom data were obtained.[9] Numerically, this sifting of evidence is drastic; but it is preceded by a quantitative, geographic analysis of the distribution of land agencies in the cutovers and a qualitative analysis of their modes of operation and land requirements. Also, this intensive analysis of the selected data is followed by a brief review of the experience of the types of agencies not otherwise analyzed. As a result, the basis of the selection and its relevance to the hypothesis are clear.

The intensive analysis is organized topically. It is a mixture of mass average and frequency summaries of quantitative data on all the settlers and by projects, of qualitative descriptions, of occasional project comparisons, conceptual analyses, and synthetic case histories. Following this presentation of evidence by topics, there is an over-all quantitative summary for all settlers and all projects; and then there is a brief case report on each of the fifteen projects.

This intensive analysis is marked by three directing ideas, all of which are obviously a development of the beginning hypothesis. (1) A settler's success is judged on the basis of acreage of land cleared and gain in net worth and, to a lesser extent, on the amount and distribution of current receipts. (2) A settler's background and experience, length of time on the place, age, and beginning net worth are factors that might be strategic aside from the activities of the colonization company. (3) Colonization projects are differentiated on the grounds of their soil location, terms of sale and sizes of units, lending and supervisory policies, improvements furnished, community developments, prices of land and costs of selling, sales methods, and other less important phases of their policies and practices.

Consequently, there are cross-classification summaries of acreage cleared—by types of settlers, by ages, and by beginning net worths for all the settlers as a whole; and each of these items is also averaged

[9] It is worth noting that in this bulletin even the introductory statements, so often given as mere background, are directly related to an explanation of the problem. Even the description of physical conditions in the region is, in respect to every item, given with a direct statement of its relevance to explaining the reasons for a retarded settlement experience in the cutovers.

for each project. The influence of time on a settler's achievement is indicated by cross-tabulating the indicators of success for all settlers by years since settlement.

Other items, such as land classification, the price of land, costs of selling land, land taxes, developmental policies, credit practices, and provision of improvements, are given only by projects.

In nearly all these parts of the analysis, there is an intermixture of summaries of the tabulations and cross-tabulations, of qualitative discussions of the points under consideration, of brief case reports and analogical comparisons among the projects, and of synthetic or composite accounts of settlers' experiences and colonization company practices.

Following this topical analysis, there is a separate section in which a brief project-by-project resumé is given. In these case reviews, there are some comparisons and contrasts among the projects, a brief statement of the factors which appear strategic in each case (in respect either to the settlers' qualifications or to the characteristics of the colonization scheme), and some qualitative or quantitative appraisal of the success achieved in establishing going farm units.

Finally there is a section in which the rest of the land sales agencies (other than the colonization companies included in the intensive analysis) are reviewed. These agencies are grouped in terms of the predominant characteristics of their policy (such as "agencies promoting the sale of lands for special uses" and "lumber companies selling land without assuming much responsibility for settlers"); and the reviews are largely qualitative statements of the practices followed and the success or difficulties encountered by such agencies.

Although a good deal of the central analysis is based on an item-by-item review and summary, still there is some exploration of the actual experiences as they occurred on the colonization projects. Such analyses of the interaction and sequence between practices and consequences in these colonization experiments exist in the case and analogical references in the topical discussions, and especially in the set of briefs on project case histories.

The "analysis of settlers' progress on individual projects" is not arranged in easily digestible form, but careful attention to it reveals that here the specific conclusions of the bulletin find their strongest substantiation. It is in this section, and only in this section, that the success of the settlers is directly tied up to the policies of the companies. Where, for example, difficult clearing conditions are offset by helpful company policies, or where excessive land prices offset other favorable factors, or where company policies have encouraged ill-equipped or well-equipped settlers, and so on, such factors are noted. There is not, however, a clear-cut and consistent analysis of these projects except that after seven case reviews the statement is made that "All the preceding seven projects have been classified as giving 'little or no aid' to the settlers." Projects VIII–XV are put in a class with aid to the settlers and some of them are called intensive projects. Beyond this classing of the projects, however, the reader has to hold in his mind the other strategic and complementary factors that are related to success or failure on each project in these case reviews. Only if the reader does this, can he see just how the conclusions of the bulletin come from the evidence which is available.

It may be noted, in comparison with Minnesota Bulletin 180, that in the U.S.D.A. Bulletin 1295 the pulling together of factors within each case is partly performed as a part of the analysis, whereas in the former bulletin the data are given so that the reader, if he wishes, may himself try to join the interactions and sequences within the units observed.

In summary, this bulletin presents a problematic situation, a formulation of the problem, and conclusions that are consonant with the characteristics of such phases of research as noted in Chapter III. Between the posing of the problem and the conclusions, there is evidence of relationships between factors for the settlers as a whole and for the project groups. But in addition to these suggestions of possible relations some of them are actually checked as they exist within each colonization project. This terminal testing of the concluding form of the hypothesis indicates a high degree of warrantability of the conclusions. It is marred, however, by a lack of clear-cut procedure for testing all of the projects — that is, by the failure to apply the same type of analysis of interaction and sequence to the experience of the settlers within each project.

WISCONSIN BULLETIN 399, 1928, AND WISCONSIN BULLETIN 406, 1929

Actually, by the time U.S.D.A. Bulletin 1295 was in print, the situation in the cutover region was undergoing marked change. The most obvious indication of the change was a tremendous wave of tax delinquency, which was experienced in the northern counties of the Lakes States. In Michigan the land economic inventory (noted above in Chapter II) as just getting under way; but such slow-moving efforts, aimed only at providing elemental facts for private decisions, were overshadowed by the public problems that mountainous tax delinquencies presented.

The University of Wisconsin issued a quick review of the facts in a bulletin issued in 1928.[10] This report, Wisconsin Bulletin 399, is not a problem-solving analysis. It includes the presentation for each of 17 northern counties of secondary figures on over-all land cover, which show that the decrease in marketable timber has swiftly outrun the increase in farm lands. The central body of the report is a quantitative presentation of the status of tax levies in these same counties over the period 1921–1927. Finally, in a section "What Shall Be Done about It?" there is a brief review of the already available instruments (school equalization, forest crop law, authority for county and national forests) that might be used to handle the tax reverted lands and to ease the burden of local public expenditures. Also, the report indicates at several points that an intensive study of the problem in one county is under way; and in fact the only data in this report that are more detailed than county totals are advance figures from that study on the relationship between tax delinquency and land quality and ownership in one town.

The following year, Wisconsin Bulletin 406, the intensive analysis referred to in Bulletin 399, was issued.[11] In terms of a full inquiry, Bulletin 399 must be regarded as a preliminary review of evidence serving the function of problem formulation. The conclusion of the first study is one general suggestion: "One thing above all is clear. It is no longer believed that the settlement and development of cutover lands can properly be left to private enterprise alone . . . some form of public ownership or control is inevitable if good results are to be obtained on a great acreage in the northern counties." This broad statement serves as the beginning hypothesis for Bulletin 406. Recognizing that the "new situation" is "one on which there is no precedent to guide," the aim of this intensive study is to make some "tentative suggestions," which would be on the order of refining the hypothesis that if public land management procedures are followed, then the northern Wisconsin counties will avoid their present difficulties of idle land resources and inadequate public finances.

The conclusions, definitely labelled as suggestive, may be regarded as an elaboration of the "if" clause of the beginning hypothesis. They include the designation of districts that should be used only for forestry and recreation, the encouragement of state or federal forest land purchases in such areas, the encouragement of industrial forestry in

[10] B. H. Hibbard, J. Swenehart, W. A. Hartman, and B. W. Allin, *Tax Delinquency in Northern Wisconsin* (Wisconsin Agr. Expt. Sta. Bull. 399, 1928).

[11] B. H. Hibbard, W. A. Hartman, and W. N. Sparhawk, *Use and Taxation of Land in Lincoln County, Wisconsin* (Wisconsin Agr. Expt. Sta. Bull. 406, 1929).

such areas under the forest crop law of the state, and the development of county forestry in those areas through the process of tax reversion. Finally they advise the withdrawal of settlers from such areas by the exchange of tracts owned by the county and located in other districts by concentrating public expenditures on improvements in such other nonforestry districts and by "something in the nature of zoning."

There are three parts to the central analysis of evidence in this report. The first is an analysis of the relationship between tax delinquency and land use. In the earlier report only brief mention of this relationship was given, for it was stated that a "program for wiser use of such [delinquent] land must be built upon more facts concerning its characteristics. This will be brought out in an intensive study."

In Bulletin 406 the main technique for analyzing the relationship between tax delinquency and other factors concerning land use is that of map comparisons. In a series of maps of the county, forty-acre parcels are shown in terms of their tax delinquency status, type of ownership, quality of forest cover, condition of terrain, and situation as operated or abandoned farms, business or recreational places, marketable or poor timber cover. There are also tables in which the county is divided into districts on the basis of the degree of tax delinquency. These districts are compared as to percentage of idle land, farm land, timber land, resort land, and abandoned farm land. "Abandoned farms, large areas of idle or unused land, and the absence of resort possibilities are all found to be closely associated with tax delinquency. Light sandy soil, stony land, rough land and swamp land are . . . generally, but not always, found associated with tax delinquency. Operated farm land and land with merchantable timber are seldom delinquent."

It should be noted that these associations or relationships are established in this case on the basis of the predominant occurrence of attributes in geographic areas. There are no data showing the sequential experience of those who owned or used the land and who are or are not paying taxes. Only in an exceptional instance is the relation explicit. At one point it is stated that "as the timber is cut, the owners are allowing this land to become tax delinquent. The large block of nontax-paying land in District III (Town of Harding) from which the timber was recently removed, is evidence of this fact. That the remaining merchantable timber land will become tax delinquent as soon as the timber is cut, is the expressed prediction of the owners of the bulk of it." In other instances, such relations are only suggested with no specific experiential evidence presented, as in the statement that "The sale of land with poor agricultural possibilities to men who have had little or no farming experience, and the policy of locating these settlers in widely scattered areas are two of the primary causes of abandonment."

The second section of the analysis is primarily an exposition of the manner in which heavy tax delinquencies in one town necessitate a rise in the tax levies on properties in other towns, and in other parts of the state. This analysis of the process of spreading tax difficulties is accomplished by a case analysis of a single town. This analysis also uncovers serious defects in the school equalization law referred to in Bulletin 399 as a "definite force" in helping to solve the idle land problem.

One other point should also be noted about this analysis of a single town. Apparently, some information was obtained in this town regarding the settlers themselves. In the bulletin this information is given in four short (and almost hidden) sentences which say that of 35 families in the town, only 3 receive most of their income from farming, 15 receive only a small part of their living from farming, and nearly half receive "substantially no income from farming." "Most of these people do little or no farming, but instead work in the woods or in near-by cities—or do not work at all." No further analysis is given. But is this not enough to raise a serious question: Are these families not farming because the soil and location make them unable to earn income from farming, or is their present location wholly unrelated to agricultural possibilities? It is clear that this question can be answered only by analysis of the sequential experience of the people involved. It is also clear that the question is important enough that different formulations of the problem (and thus of hypotheses, analyses, and conclusions) would depend on the answer. But nowhere in the literature is there indication that the "commercial agriculture vs. forest" formulation of the cutover problem was seriously questioned in research until 1938.

The next section is entitled "Forestry Possibilities." The data presented are acreage figures on types of forest cover and on the ownership of forest land by farmers, lumber companies, land companies, and individual nonresidents. On the ground that the bulk of tax delinquencies occurs on land belonging to land companies and nonresident individuals, and that ownership stability is essential to forestry management, the remainder of the section is devoted to a conceptual analysis of how the county should proceed to develop public forests. Finally, under the heading, "County Forests Will Prove Profitable," data on present forest stand are projected into yield forecasts, and these are converted into gross annual revenues per acre — which are calculated to be five cents per acre for a decade, then rise to 60 cents in 40 years, and one or two dollars per acre per year thereafter.

Insofar as this study deals with the problem of instituting action to handle lands already tax delinquent, there was no opportunity to test its recommendations against experience, locally at least. In previous studies, undesired consequences appeared in sufficient number to create a problematic situation; but the elements in the problem at the time were such that recommendations could be made for the institution of action to direct the outcomes of experience along those lines of action which had at the time already shown up as successful. In the later instance, however, the central problem was not how to get more experiences to come out like the successful ones; it was what to do with an entirely new pattern of affairs.

The 1929 study might have taken up the situation of the taxpaying land owners in order to search out possible means of preventing their lands from falling idle as a result of land-tax delinquency. But at the time the existence of already idle and delinquent lands was so great as to constitute "an emergency which calls for quick action." For this specific problem, the local research could do no more than arrive at very tentative suggestions — not even a penultimate test could be made since there were no results of actual experience with such suggestions. But with this limitation, a more thorough attempt at synthetic projection of the actions and consequences (beyond the superficial calculation of forest land returns) might well have been undertaken.

In sum, the inquiry falls short in respect to developing the "then" clause of the hypothesis, especially in respect to the alleviation of public finance difficulties. It is strong in stating what actions might be instituted but is vague as to postulated consequences. Furthermore, had a direct analysis of the processes of experience by which lands had actually passed from ownership or occupancy to no use and delinquency, by which some nonoccupied lands had remained nondelinquent, and by which some lands still occupied had gone delinquent, there would be greater assurance as to how the suggested actions would fit into the ongoing experiences of the people involved. And there is reason to believe that more attention to the actual experiences of the people on the land might have, in turn, suggested a revision of some of the ideas which dominated the formulation of the problem and perhaps affected the conclusions.

U.S.D.A. TECHNICAL BULLETIN 92, 1929[12]

Without pursuing an elaborate review, one or two points should be made in connection with an economic study made by the U.S. Forest Service in the Michigan cutover area. This is a study of "forest destruction," which is seen as "a *process* of harvesting and manufacturing the raw materials provided by nature" without concern for "the production of more raw materials to take the place of those consumed."

In the introduction, it is stated that the study "was undertaken for the purpose of ascertaining how forest destruction, followed by nonutilization of the cutover land, affects the regions concerned. Michigan was chosen because it is one of the States in which the more serious forms of destruction had the earliest start. . . . It affords one of the best examples." But the very next sentence reads as follows: "*That the effects on the region of such conversion are fairly obvious* is indicated in the words of the State tax commission," and then there follows a long paragraph taken from a state report dated ten years before the bulletin in which the quotation is made.[13]

Why should a research study be directed to determining the consequences of a process if, ten years before, these consequences were already fairly obvious? The answer is given two paragraphs later: "A vastly broader program [than any yet attempted] will be required if the situation is to be adequately met." In other words, the research objective is not only to recount the process by which the cutting of timber leads to undesirable consequences; it is to find modes of action which, if instituted, would prevent the practice of cutting timber from resulting in these "fairly obvious" consequences, and to suggest ways of dealing with lands already denuded. In other words, the purpose is to search for means to unify actions and consequences.

The bulk of the bulletin is devoted to a thorough amalgamation of secondary data (chiefly from the census and governmental reports) to describe the history of timber removal in Michigan, the progress of population growth and agricultural expansion, and the effects of the stoppage of lumbering and sawmill operations on employment, local farm markets, tax revenues, railroad service, tax assessments, and so forth. It should be pointed out that some of these effects, such as the decline in sawmill employment and the abandonment of lumber-hauling railroads, are clearly related over time to the decline of lumbering. Other effects, however, are only associated relationships

[12] W. N. Sparhawk and W. D. Brush, *The Economic Aspects of Forest Destruction in Northern Michigan* (U.S.D.A. Tech. Bull. 92, 1929).

[13] Italics inserted.

and are static. Examples are the rate of interest on farm mortgages or the miles of road per county. A section on "The Problem of Idle Land" presents data showing, as did Wisconsin Bulletin 399, that agricultural expansion has not proceeded fast enough to make use of any large part of the cutover lands.

The last part of the report is given to the presentation of "the major features of a program for public action." These recommendations vary from national forest acquisition to publicly subsidized credit for private forestry operations. In this bulletin, as in Wisconsin Bulletin 406, at least a large part of the problem dealt with was such that locally, at least, it was not possible to find ex post facto experiments as tests of the recommended actions. This is clear from the fact that only 2 per cent of the wooded area was already in public ownership. But the evidence of persistently negative consequences from private forest operations in the past is strong enough to give these recommendations highly suggestive force.

By the same token, however, the suggestion in respect to the greater acreage that is recommended to remain in private hands has much less force. The evidence emphasizes that lumbering, not forestry, has been the dominant motive of forest land owners. In this situation, a careful study of the owners would be necessary before a proposed program designed to tempt forest land owners to practice forestry would have even strong suggestive force. The report does mention that some forest land owners operate wood-using industries and that they, in contrast to the lumbermen owners, would be interested in increasing their future timber reserves. But this brief reference is wholly inadequate in view of the importance of the potentialities of private ownership to the suggested program of action.

U.S.D.A. CIRCULAR 160, 1931 [14]

In 1931 the U.S. Department of Agriculture issued a bulletin that summarizes all the settlers' records secured in 1919–1920 (some of which had been used in U.S.D.A. Bulletin 1295, issued in 1925). It also contains information obtained from a rapid survey in 1928 of the same areas visited in 1919 and 1920.

The bulletin is offered as one that "describes and presents some of the fundamental conditions affecting the development of this cutover region and summarizes briefly the progress that settlers may expect to make in developing cutover land farms." The specific conclusion of the report is "that until a directed settlement policy based

[14] W. A. Hartman and J. D. Black, *Economic Aspects of Land Settlement in the Cutover Region of the Great Lakes States* (U.S.D.A. Circular 160, 1931).

on an economic classification of land is established, the prospective settler can eliminate certain hazards only by exploring all reliable sources of information pertaining to the area in which he is interested, and by weighing the advantages and disadvantages."

After a lengthy description of the cutover region, based on physiographical maps and previous studies, the report summarizes all the 1919–1920 records of over 2000 settlers in the three states by averages of the records grouped, by two-year intervals, as to length of settlement and subgrouped as poor, good, and average settlers. The presentation summarizes the information on the group of average settlers in their progress from the first two years of settlement through their twentieth to twenty-first years. Clearly, three comments are in order as to the inadequacies of this analysis. First, since the early part of the report explains the turnabout in the agricultural situation after World War I, these data could be of only limited value to describe the progress of settlers in terms relevant to 1931. Secondly, especially in view of the rising farm incomes and land prices from 1910 to 1920, it is a dangerous procedure to describe the one- to two-year settlers who actually settled in 1918–19 as though their record were comparable to the first two years in the settlement experience of the veteran settlers whose first two years were experienced before 1900. Third, since the 1925 report made clear the wide variations among certain of the colonization areas, the lumping together of settlers from areas that are widely scattered and that have been shown previously to differ in respect to certain strategic factors is obviously questionable. In short, this whole analysis constructs a quantitative process of settlement that no one has lived through.

In the third section, a brief qualitative summary is given of the history of settlement in the region, by settlement areas or groups of areas. These case histories and groups of histories are based on notes from a rapid field visit or a letter from the local tax assessor, or both. No direct individual settler data were obtained, but one or two items in the 1920 data were partially summarized by dividing the farms as to whether or not they had changed ownership by 1928. In each of these summaries at least a cursory search for possibly strategic factors which might explain the 1920–1928 abandonment history is sought. Most important, however, is the fact that in the areas with extremely high percentages of foreign-born settlers, only an ordinate amount of abandonment or foreclosure is found — even though some of these areas are described as of very poor soil quality. These "observations suggest that nativity of settlers is of primary importance," although "it would be necessary to analyze many contributing factors." Clearly, on the evidence used in this report, it would appear that this point is a high light of the study, at least as a highly suggestive hypothesis for further work. Again, as is the case with U.S.D.A. Bulletin 1295, the unit history summaries are the basis for such insights as the report is found to produce. Unfortunately, however, the handling of these unit histories and of the settler types within them is not sufficiently consistent to provide assuring comparisons and contrasts of the settlers' and the projects' experiences. Yet, informal and brief as these area histories are, they represent the only attempt to present the story of what happened in some parts of the cutover in the 1920's; for, in contrast, Wisconsin Bulletins 399 and 406 reported facts that were indexes of results, not examinations of the process that led to such results. Still, it must be remembered that this 1931 report did not deal with land that never was settled, or with land that was settled after 1920, but only with land on which there were settlers in 1919 or 1920.

In sum, this study lacks a formulated problem, a directing hypothesis, or a significant conclusion — either as a recommendation for action or as a hypothesis for further research. In contrast to the bulk of the materials which merely repeats general information previously known or which is so outdated and inverted as to be of dubious value, the brief presentation of histories of settlement projects (though inadequately documented and loosely organized) gives some evidence of obscured but potentially important suggestions for reappraising past analyses of the land settlement process.

THE MINNESOTA REPORTS, 1934–1935

Seldom have the suggestive conclusions of social science research been put to test in experience more quickly or more widely than were those of Wisconsin Bulletin 406.

The slow process of a land inventory which covered over four million acres of land in Michigan between 1922 and 1929[15] and one Wisconsin county between 1927 and 1929 was not enough if the suggestions of Bulletin 406 were to be used. Accordingly, in 1929 the Wisconsin Agricultural Extension Service launched a series of "emergency land surveys." For these, technical workers assembled maps comparable to those in Bulletin 406 and, together with committees of local citizens and public administrators, they demarcated forestry and farming zones for various counties. By this process the recom-

[15] Wade DeVries, "Michigan Land Economic Survey," *Journal of Farm Economics*, vol. X, 4 (Oct. 1928).

mendations of Bulletin 406 were repeated and made specific for five cutover counties between 1929 and 1931.[16]

Beginning in 1932, things moved fast. By midyear of 1933, one county had put the police power behind its forestry-farming map by the use of a zoning ordinance. And in rapid succession 25 other Wisconsin counties followed suit. Meanwhile, counties took tax titles, entered the land for the benefits of the forest crop law, and exchanged lands. Federal relief programs came into existence in 1934 to extend public ownership in the forestry areas and to help move some settlers out of restricted districts.

In 1932 the Michigan program turned from inventory work to land use planning, i.e., the making of recommendations as to the specific direction of public efforts to affect land utilization.[17] Meanwhile, Minnesota had in 1929 started a very complete land economic survey of one county.[18]

Even a reading of the first page of the report of the Minnesota land economic inventory on Hubbard County, published in 1935, gives evidence as to why the inventory procedure was too slow. The record indicates that after two annual $25,000 legislative grants, a third grant of only $6000 was given, not to continue inventory work for "all lands in the State" as originally intended, but to complete and publish the work previously undertaken. This report of 264 pages and two elaborate colored maps represents a compendium of detailed facts on the soils, land cover, land ownership, land use, tree species and types, school enrollment, tax values, tax rates, delinquent taxes, mortgages, and population for the entire county. No data were secured by direct interview of the people who owned, operated, or lived on the land in the county. For the most part the data on each topic are summarized separately and the main emphasis is on comparing the figures among the townships (or school districts in the case of educational data).

[16] Extension Service, University of Wisconsin College of Agriculture, *Making the Most of Marinette County Land* (Special Circular, 1929). With the same title, see *Ashland County* (1930); *Forest County, Oneida County, and Taylor County* (1931); *Washburn County* (1933). Also see *Langlade County, A Survey of Its Natural Resources and Their Utilization* (1934); and *Making the Best Use of Wisconsin Land Through Zoning* (1934). A bulletin on *Recreation as a Land Use* by G. S. Wehrwein and K. H. Parsons (Wisconsin Agr. Expt. Sta. Bull. 422) represents a preliminary opening of this topic by counting the properties in, acreage of, and tax revenue from recreational lands in 3 cutover counties.

[17] L. R. Schoenmann, "Planned Land Use," *Proceedings*, American Association for the Advancement of Science, June 21, 1932, pp. 48–57; and "Land Inventory as a Basis for Planning Land Utilization," *Proceedings of the National Conference on Land Utilization*, Chicago, November 19–21, 1931 (Washington, 1932).

[18] *Land Economic Survey of Hubbard County, Minnesota* (Minnesota Agr. Expt. Sta. Bull. 317, 1935).

The recommendations of the study call for the state to keep lands that it already owns, the adoption of rural zoning, relocation of settlers from poor agricultural lands, public retention of the delinquent lands, extension of public forest lands, promotion of recreational attractions, and changes in local government organization. These recommendations, substantially the same as those of Wisconsin Bulletins 399 and 406, are no more convincingly substantiated, in spite of the mountainous pile of data contained in the report in which they are found. Nor is it made clear what the institution of these actions would do for the people of the area or how the action could be locally applied. Rather, the achievement of the project must be accepted in the terms in which its object is stated: "to assemble information on the various factors involved in the utilization of the natural resources of the area studied that will serve as a guide in determining upon the best program of utilization of these resources." "The report in itself cannot, of course, do these things; it can only supply the data."

The unwieldiness of the inventory procedure in relation to its usefulness in providing solutions to problems was amply demonstrated in Minnesota, about the same time that the Hubbard County report was printed, by the publication of two books dealing with the entire region.

In 1934, a Minnesota governor's committee published a book directed toward obtaining greater public interest in the seriousness of the situation in the northern sections of the state.[19] The recommendations of the committee follow the pattern previously given. The supporting evidence consists mainly of a series of statistical comparisons between northern and southern counties in respect to physical, financial, social, agricultural, and governmental conditions.

The next year another book covering the entire Minnesota cutover region was issued.[20] Whereas the previous book did not involve any independent survey, the second does offer some new materials. So far as analysis of data showing the existence of problematic conditions in the cutover area is concerned, this book is like the former one. Also, the recommendations for action are essentially the same. But there is an important difference between this study and the preceding reports or studies of the area — a difference that is great enough to

[19] Committee on Land Utilization, *Land Utilization in Minnesota* (Minneapolis, 1934). See also R. G. Blakey and Associates, *Taxation in Minnesota* (Minneapolis, 1932), especially Chap. 5, "Tax Delinquency and the Cutover Land Problem in Northern Minnesota."

[20] O. B. Jesness, R. I. Nowell, and Associates, *A Program for Land Use in Northern Minnesota* (Minneapolis, 1935). Actually, the research work for this book was already under way when the governor's committee was appointed.

justify the subtitle of the book, *A Type Study in Land Utilization.* The difference is that in the section dealing with proposed policies and programs, there are what may be called synthetic or projectional analyses of how the recommendations might work out in practice. Under land classification, for example, the study embodies both the type of work of Wisconsin Bulletin 406, in which rapid demarcation procedures were worked out, and of the Wisconsin Extension series in which the actual divisions between forestry and farming zones were made for all the problem counties.

Other and more original examples are the calculation of incomes to be expected by promoting a program of private forest land management, the determination of areas of forest land suited for public ownership, the calculation of costs and savings of relocating farm families, and the analysis of possible savings from governmental reorganization and consolidation.[21]

In reference to earlier studies, it was pointed out that in the absence of any experience in which proposed means and postulated consequences were known to have been unified, a research study could hardly go beyond a very tentatively suggestive stage. But it might go as far as to check the hypothesis conceptually by symbolic or synthetic projection. It is in this step that the second Minnesota book is notably different from the previous analyses.

Essentially, then, the 1935 book, so far as the advancement of scientific research is concerned, may be said to have given specificity to an enlarged hypothesis — specificity in respect both to the means to be instituted (what and where) and to the consequences to be expected.

Yet it should be noted that in spite of all the research reports that appeared between 1925 and 1935, only scant information is revealed in any of them as to the direct history of experience of the families in the region or of the outside owners of land in the region.[22] Practically all the inquiries were based on the treatment of data which were only suggestive of "who did what and why."

[21] The synthetic method had previously been used in farm management work and on a small scale in analyzing the results of reorganizing local government. See G. S. Wehrwein and B. Allin, *Possible Farm Tax Reductions through Changes in Local Government* (Wisconsin Agr. Expt. Sta. Special Bull., 1933); M. M. Regan, "Possible Savings in the Cost of Governmental Services Arising from the Relocation of an Isolated Community," *Minnesota Farm Business Notes,* No. 143 (1934); G. S. Wehrwein, "Problems for Research in Public Finance Arising from Land-Use Zoning Programs," *Journal of Farm Economics,* vol. XVI, 1 (Jan. 1934). Cf. P. A. Eke, "Synthetic Method and the Principle of Comparative Advantage in Land Utilization and Farm Management Studies," *Journal of Farm Economics,* vol. XIV, 2 (April 1932); and H. R. Tolley, "Recent Developments in Research Method and Procedures in Agricultural Economics," *Journal of Farm Economics,* vol. XII, 2 (April 1930).

[22] Cf. with the reference to this same point in the review of Wisconsin Bull. 406, above.

The importance of this fact is illustrated by the criticisms of the Wisconsin area classification system made in the 1935 Minnesota report. It is stated that the Wisconsin system, in emphasizing tax delinquency as an indicator of problem areas, overlooks the possibilities that some unsuspecting absentee owners pay taxes on very poor lands, that some taxes are paid on the same kind of land only in anticipation of timber stripping, and that some owners of good lands are voluntarily delinquent in tax payments. The important point of this criticism for our purposes is that it reveals that any static attribute of a person or a piece of land cannot give warranted evidence as to the actual process or history through which that person or piece of land is moving. In fact, the closest that any study published between 1920 and 1935 came to revealing such evidence is to be found in the brief stories of settlement projects reported in U.S.D.A. Bulletin 1295 and Circular 160. And as has been pointed out, these analyses were not only incomplete in themselves, but also they dealt only with lands on which there were settlers in 1919–1920.

The 1935 Minnesota report makes one other brief informal reference to a phenomenon in the cutover area quite at variance with any that had been otherwise reported—one completely opposite (because of the passage of time) to the 1920–1928 processes reported in U.S.D.A. Circular 160. This point deals with the movement of urban unemployed into cutover land beginning with the start of the Great Depression in 1929. The trek of unemployed from the cities to cheap cutover hide-outs was the force that had actually stimulated citizens in Oneida County, Wisconsin, to pass the first rural zoning ordinance in 1932. But this phenomenon did not reflect itself in any significant way in the research work done in the region during the period 1932 to 1938 (and even later in many cases) when attention was being given chiefly to putting into action the solutions originally suggested in 1928 and 1929.[23]

THE SOCIOLOGISTS' REPORT, 1937–1938

In 1937 and 1938 two research reports on northern Lakes States studies were made by rural sociologists at Minnesota and Wisconsin. Both of these studies have to do with projects set up by federal agencies to purchase land in certain areas in order to remove the settlers from them, and then to help some of these people resettle in relocation areas.

[23] There were some farm management studies reported for the region during this period, but by that time the differences between farm management and land utilization research purposes were sufficiently wide apart to be regarded as separate here.

The Minnesota bulletin is essentially a report on the development of one of these resettlement projects between 1934 and 1937 [24] The study does not pretend to be of a problem-solving type. Rather, inasmuch as the resettlement project is regarded as a social experiment, the research work was undertaken in order to provide "a record of conditions existing when the project was started and a clear statement of improvements hoped for" in order to "serve as a very valuable basis for a future study five or ten years hence, when the actual results of this or similar projects may be better determined by experience."

Such an objective for a preliminary to a full inquiry certainly meets the requirements of experimental social science research expressed in Chapter III. In view of the limited, but important and carefully stated, research objective of this publication, very little can be said of it. It should be noted, however, that the information with respect to the people of the area is presented in gross terms—averages or totals—for the communities in which they lived in 1934. Particularly in view of the fact that the report stresses the scattered location of the new homesites chosen by the families, it is clear that much of the data will not be directly useful for a future analysis of the outcome of the experiment. Since the old communities have been permanently depopulated by the experiment and since the "final test" of it will be in the human adjustments it will bring, any future, meaningful grouping of the subjects of the experiment will obviously have to be made on some other basis than the particular locality within the purchase area from which they were removed.

It should also be noted that in this report, specific mention is made of the "new families" in the area in 1934 who had been "attracted by cheap or free land . . . especially since the depression, replacing part of the population" which had left in the preceding decade.[25]

Another bulletin published in Wisconsin in 1938 also dealt with

[24] R. W. Murchie and C. R. Wasson, *Beltrami Island, Minnesota* (Minnesota Agr. Expt. Sta. Bull. 334, 1937). Mention should also be made of two minor reports issued in 1937 both of which present data on the extent of isolation of settlers in certain parts of the cutover region: G. S. Wehrwein and J. A. Baker, "The Cost of Isolated Settlement in Northern Wisconsin," *Rural Sociology,* vol. II, 2 (Sept. 1937); and Minnesota Land Use Planning Staff, *Isolated Settlement and Tax Delinquent Land in Northern Minnesota* (U.S. Resettlement Administration Land Use Planning Pub. 12, 1937).

[25] In this respect also note the suggestive statistic in a national report in 1935 which showed that, in 1933, in two sample cutover counties 31 per cent of the rural nonrelief male heads of families and 45 per cent of the relief male heads reported nonagricultural activities as their usual occupation. T. C. McCormick, *Comparative Study of Rural Relief and Non-Relief Households* (U.S. WPA Res. Mono. II, 1936), Tables T and U, pp. 53, 55. Also see footnote 22 above.

an existing relocation project.[26] But here the purpose of the study was "to determine the social problems involved in a land retirement and resettlement program" and perhaps to "offer some suggestions in formulating future policies for their solution." The report makes clear that there was doubt as to whether a program "designed to retire from cultivation those tracts of land which are poorly adapted to arable farming" was in "agreement" with "the improvement of the human factors," i.e., "the welfare of the families involved."

The report, based on schedules taken from all families in the purchase area and from a sample of families regularly living in the resettlement area, is composed of three parts. In the first, all the information from the schedules is presented topically, as totals, averages, or percentages for all the families in the purchase area. These figures are compared with those for all the sample families in the resettlement area. All through these comparisons of the two areas, statements are made implying that project supervisors might not give adequate attention to the "economic, social, and culture patterns." Even were the differences revealed as significant in the gross as is implied in the written summaries, the basis of the comparisons is severely shaken by an admission at one point that part of the variation between the areas in respect to a certain item is "due to the much higher proportion of single-member families, especially of the extreme aged" in one of the two areas. And shortly thereafter, it is pointed out that "In fact, average figures for the Purchase area as a whole are misleading, because of the influence of" two communities "which are so much different from the other Purchase Area families in almost every respect." One can only wonder what purpose is served by twenty pages of such area-average comparisons.

The second part of the report consists of four analyses of data for two neighborhoods, a group of scattered settlers in the purchase area, and one neighborhood in the resettlement area. Here again the analysis consists of comparisons of average figures for the families in each group with the averages for all families in the two larger areas. In addition, there is a comparison of the neighborhoods based on maps showing, by lines, the extent to which the families visit with friends, exchange work, or obtain emergency aid from their neighbors.

The point in these comparisons is that within the purchase area there are closely-knit neighborhoods, which enjoy a comparatively good living, and other groups with low degrees of neighborliness, which have a very low level of living; while in the resettlement area

[26] G. W. Hill, W. Slocum, and R. O. Hill, *Man-Land Adjustment* (Wisconsin Agr. Expt. Sta. Res. Bull. 134, 1938).

there is a high degree of neighborliness and a level of living far above that of even the best neighborhood group of the purchase area. The clear implication is that the families most likely to succeed when moved into the resettlement area do not need to be moved and would be torn away from a successful neighborhood if they were; whereas those whose local roots are shallow would even more certainly be misfits in the resettlement area pattern of living.

This analysis is weak on three grounds. First, the comparisons are not convincing except that they show some groups to have a higher degree of poverty than others. Many of them are obviously irrelevant or picayune. But even in respect to the chief criterion used, the average value of consumption is $650 for the neighborhood with the "highest degree of social life" and "solidarity" and is $651 for the entire purchase area. Secondly, the interpretations of the data are not always clear. Thus, at one point, the sale of pulp, stove wood, berries, and moss is interpreted to show that farmers "lacked either the desire or the necessary capital to farm," but in the case of the highly integrated neighborhood, such sales show that "All resources of the locality were drawn upon to produce a living" for these "thrifty, frugal folk." Thirdly, the mapped patterns of social contacts show a good deal of neighborliness among 56 families in a strictly rural area approximately 35 square miles in size. In comparison with this situation, another area is designated as unneighborly because 16 families in a space of 4 square miles visit more with friends in a small city less than 3 miles distant than they do with each other. If this analysis suggests that there are marked differences in the experiences of the families in the two areas, it is not because of any evidence of more or less sociability on the part of the families concerned, but because one group is more suburban than the other—yet this aspect is not mentioned. Furthermore, there is nowhere evidence or even a conceptual argument given as to why "lack of interaction of families with one another" would partly "account for the difference between success and failure."

Some of the foregoing comments might be interpreted as indicating poor judgment rather than a deficiency in basic research method or approach. The reason for the foregoing analysis, however, is to contrast the first parts of Wisconsin Bulletin 139 with the last part, in order to indicate that the difficulties noted are not only matters of judgment in interpretation but are unnecessarily accentuated by the whole approach taken to the real problem of the inquiry.

After erroneously claiming that the preceding discussion shows that "the factors which have brought them [families in the poorest groups] to their present low standards were often psychological and social in nature . . .," the report correctly continues: "However, the hypothesis that the welfare of Purchase Area families will be improved by moving them into the Resettlement Area cannot be tested merely by making a comparison of the standards existent in the two areas." The question then raised, which it would seem is the research question for the whole study, is: "Do the plans make adequate provision for all families?"

Following this question, the report notes that the project will extend resettlement aid in three ways: to those who meet a set of qualifications for full-time farming, part-time farming, or retirement homesteads. The families in the purchase area are then grouped according to whether the family meets the set of qualifications for any of the three types of aid. It is found that 85 families fit into these three groups and a discussion of each group is given. Then, it is revealed that 62 families do not fit into the proposed plans at all. The analysis of these families is again presented largely in terms of group averages, but it is important to note that the groups are relevant to the problem of resettlement: 31 aged families, 16 broken and single-member families, and 15 nonfarm families. It is clear that this classification is not logical in the sense of classifying aged and young or nonfarm and farm families; it is, however, logical in reference to the character of the problem under review. The main suggestion that might be advanced is that instead of giving the analysis in terms of averages for even these groups, further subclasses would be more revealing. For example, once it is established that 10 of the 16 in the second group are persons living alone, there is little point in presenting an average consumption figure that includes them along with six one-parent families with young children. What is more, it is obvious that the relocation problems of these two subgroups are sufficiently different to require separate treatment in the solution of those problems.

The lesson to be drawn from this bulletin is clear. It may be put in the form of a question: Why offer masses of comparisons of questionable statistical average data and then in a final chapter just begin to break the data down into parts that are revealing of the human experience of living families and are also relevant to the formulated problem and the hypothesis? The chief conclusion of the report—that plans need to be modified if they are to serve the welfare needs of all the families in the area—actually is based on the final 13 pages (out of 70) in the analytical part of the bulletin.[27] The refinements

[27] In fact, the only conclusion that does not depend on these few pages is one suggesting that the plans should "consider" existing neighborhood groupings. The weakness

of this substantiated general hypothesis are limited by the degree to which this six-page analysis breaks down the relevant categories of the data. Had the entire inquiry been structured by these categories of facts which comprise evidence for the problem at hand, not only would a good deal of useless and doubtful material have been eliminated (thereby increasing the effectiveness of the whole report), but also further refinements of the hypothesis could have been tested so that additional specific suggestions for the conduct of the project might have been forthcoming.

U.S.D.A. TECHNICAL BULLETIN 687, 1939 [28]

In 1939 the Bureau of Agricultural Economics reported on an inquiry, begun in 1936, which was designed as an "analysis of land use problems" in Forest County, Wisconsin in order to develop a plan for that county. This report illustrates the point made in Chapter II that with the development of rural zoning, submarginal land purchase, settler relocation, and other programs after 1935, almost the whole effort of land economists turned to planning land use.

This bulletin is also illustrative of two other important points. First, it indicates how far the problem of tax delinquency had turned attention away from direct analysis of the experience of people on the land and the extent to which research attention centered on problems of local government. Nearly half the space in the report is devoted to analyses of tax data and revenues and expenses of government; and the criterion used to indicate consequences is savings in governmental expenditures. In contrast, the only direct information on the experience of the people in the area is a five-paragraph summary of farm incomes, a few lines on population trends, and a brief running description of areas in the county, given in connection with an area classification map.

Secondly, in contrast to the viewpoint of the two bulletins just reviewed, in this report, public action programs are not regarded as experiments available to test the hypotheses and problem formulations which led to their adoption. Rather, these programs are cemented into a fixed problem formulation; and the inquiry proceeds only within that framework. Such inquiry serves a necessary and

of the neighborhood analysis has been shown above. Also, be it noted, that Minnesota Bulletin 334 emphasizes that when resettlement families were given complete freedom in selecting their new homesites, old neighbors voluntarily chose to scatter in all directions.

[28] V. W. Johnson, S. Henderson, and J. H. Marshall, *A Land Program for Forest County, Wisconsin* (U.S.D.A. Tech. Bull. 687, 1939). Also see W. F. Musbach, *Land Classification for Land Use Planning in the Great Lakes Cutover Region as Illustrated by Forest County, Wisconsin* (U.S.D.A. Land Economics Report No. 1, 1937).

useful function in either public or private action. But it is important to understand then that the area of inquiry is restricted and that scientific investigation is being applied, not to the larger problematic situation for which the action programs may have been devised, but only to the smaller field of administrative decision.

In the present case, for example, the problematic situation as briefed in the opening sentence is a very broad one: "Many cutover counties are confronted with a complex of land use and local governmental problems that are holding down the income of their citizens and seriously threatening their continuation as solvent units of government. Such problems exist . . . in Forest County, Wisconsin."

For this problematic situation, the formulation of the problem as given in the same and following paragraph reads like a conclusion: "The problems arise fundamentally from maladjustments of the population to the resources of the county. But certain land programs and policies have been established which . . . will grow in importance as measures for the betterment of the welfare of the people." But this statement, if a conclusion, is a conclusion of previous inquiry, not of the present inquiry; for the programs referred to are not regarded as experimental actions with respect to the stated consequences desired: higher incomes for the citizens and solvency of local government.

The point that has arisen in the reviews of several of the bulletins in regard to the effects of urban unemployment may be used here as a testing instrument. Early in the bulletin, attention is called to the importance of distinguishing among farms listed in the census those that are farms and properties of people "who live in the open country but who do little or no farming." Later, it is stated that often "the location of farms has been determined by proximity to temporary nonfarm employment rather than by the productivity of the land or nearness to markets and neighbors." This same group of nonfarming families is referred to again as one that increased after 1930.

In spite of these seemingly significant observations, the only discussion of the incomes of the people in the county deals with gross farm income. In this analysis, 23 farm income records were set aside out of 167 in order to eliminate nonfarmers. This sorting was done by segregating those with less than one acre of cultivated land. Of the 144 farms, 34 per cent are reported to have no cash farm income and 31 per cent less than $100 gross cash farm income. Do these figures indicate that some of these places should be "retired from agricultural use" and that "Too small an acreage of cleared land per farm is undoubtedly one of the chief reasons for low farm incomes"? Or, taken with the earlier statements on the landward movement of unemployed, do they indicate, at least in part, that the problems may arise from sources other than a maladjustment of agricultural or forest resources to the people on these places?

Discussing the relief burden, the report states that "The chief causes of this excessive relief burden, aside from the general depression, undoubtedly are attributable to unemployment resulting from the decline of the forest industry, and to low farm incomes . . . on poor land, or to . . . inability . . . to clear enough land."

In the absence of direct evidence on experiences of the people involved, there is no testing of this type of statement. Evidence such as the decline in the timber tax base is inconclusive insofar as the decrease is admittedly owing not only to the cutting of timber but also to purchase by the Forest Service ("the largest owner of forest land in the county") and to county acquisitions through tax reversion. And evidence of the existence of nonfarm places and a 20 per cent increase in the number of so-called farms in the county between 1930 to 1935 only casts doubt on the interpretation given.

One consequence of this fixity in the formulation of the problem is seen in the recommendation that a sound relocation program "would go far . . . by placing families now on poor land in better locations where their incomes would be increased." It is further accentuated by the fact that already some 100 families had been relocated by the time of the inquiry. But this clearly experimental experience is not brought into the analysis as a test except for the statement that about half of these families moved into villages or towns, and one half moved to other farms—although it "is not known . . . how many . . . can be considered farms and how many are . . . 'rural nonfarm homes.'"

The same type of consequences can be found in other parts of this study in respect to rural zoning, governmental reorganization, and national forest ownership. While there are indications that these measures may not increase incomes or put the local governments in better financial shape, such experience as there has been with these actions is not drawn upon as evidence to test the results against postulated consequences.

MINNESOTA, WISCONSIN, MICHIGAN, 1840–1941

In 1940 one short study of a cutover problem appeared in which a land economist actually looked directly upon the experience of some people in the cutover.[29] In this inquiry, attention centered on a spe-

[29] J. E. Mason, *Isolated Settlement in Koochiching County, Minnesota* (U.S.D.A. Mimeo., LE-8, 1940).

cific source of doubt and confusion in the problem complex of the cutover. Because certain local officials did not feel that all isolated families in one Minnesota county were "causing undue expense" or "using the land contrary to public interest," an inquiry was made to determine a policy in respect to the possible need for relocating isolated settlers. Data on some 70 homesteads were secured by interview, and other data were obtained from available records of federal relocation clients.

For our purposes here, it is to be noted that in this inquiry which involved some analysis of direct experiential data, the report emphasizes not only the "interesting fact, not generally known outside of the cutover area," that "an unusually high percentage of the population in the sparsely settled sections is made up of unmarried men, living alone." Also, the data reveal so many inhabited places that are clearly not farms in any sense, that one of the main divisions made is between farmers and nonfarmers.

A second point to be noted is that after the analysis, which is mainly structured as an item-by-item comparison of farmers and nonfarmers, families and bachelors, there is a presentation of five abbreviated case histories. These, however, are given as notes and are not consistently analyzed so that they are a part of the analytical construction.

Finally, the leading conclusion of the report is the presentation of five criteria for judging the priorities by which relocation should be instituted. It is not at all clear how these criteria are conclusions of the study. If these criteria had been used as beginning hypotheses, the cases on which data were obtained could have been meaningfully classified according to these suggested criteria, and then analysis would have tested what would have been the outcome if these, or modified, criteria were instituted in directing action. Without a hypothesis relevant to the specific problem and used as a directing idea in the analysis, facts are piled up that substantiate the idea that isolation and relocation pose problems but do not test the suggested actions.

In 1940 a Wisconsin bulletin[30] presented the results of another study by sociologists which was based on an analysis of data secured from field work in a cutover county. This inquiry offers a formulation of the problem: "Problems have been exaggerated and oversimplified . . . In consequence, measures advocated . . . have been both extreme and broad-sweeping. One of the most publicized has urged

[30] G. W. Hill and R. A. Smith, *Man in the "Cutover"* (Wisconsin Agr. Expt. Sta. Res. Bull. 139, 1941).

extensive resettlement of the inhabitants on better farm land. Proposals of this type are based on the premise that poor land makes poor people."

The reformulation posed in this study is that problems in the cutover area arise out of the existence of a group of relatively incompetent people attracted to the area because they cannot compete elsewhere; but that there also is a group of competent persons who, because of preference or lack of capital, buy cheap land and make a go of it.

The hypothesis of the report is somewhat differently stated at different places, but so far as problem solutions are concerned, it appears to be simply that if differences in the competence and ambition of people were recognized, then different programs of assistance would be needed. Obviously, this is a limited hypothesis; its test cannot result in an indication of what to do to get what results. This limitation is clear from the first statement of the purpose of the study: "to measure the influence of each of the composites of the human factor on the farming enterprise."

At another point a reference is made to the substantiation of the "original hypothesis": "the nearer we come to the human side of the family-farm relationship, the more positive is our evidence that man usually chooses to make the adjustments which his desired status require." Later, "we bring our social hypotheses to their ultimate test. If the two groups (relief and nonrelief) are two distinct economic classes, and if social behavior is related to economic status, we should be able to observe differences between the two groups."

Aside from 27 pages in which three communities are described, the primary organization of the analysis is a comparison of average and percentage figures for two groups of families, those on relief and those not on relief. In addition, the relief group is divided into chronic, opportunistic, and emergency relief cases.

In the comparison of these groups, the inquiry shows that average indexes support the classification that has been made. The relief group averages show bigger families, more broken families, more unemployables, lower indexes of interests, effort, technical farm knowledge, and character, greater loss in tenure status, a greater tendency to give up farming, less social participation. With reference to only a few items are the three subgroups of relief cases used.

Although in the early parts of the report stress is laid on the fact that only half the relief families have any farming activities or intentions, and that an important influx of relief families occurred during the depression, these apparently significant groups are not held together to make it possible to see how the experiences of these people fit into the situation under analysis. In fact, this bulletin illustrates a point in respect to grouping data that has been noted in other studies. After a long analysis structured on a set division of the schedules collected, the most important conclusion of the inquiry is that a "three-fold grouping of the 'cutover' farming population must be recognized in policy formulations." But these three groups specifically are *not* the groups distinguished in the analysis; and the analysis and the conclusions so describe these categories that it is impossible to set them in any understandable order.

In other words, despite an imposing array of investigational materials, the conclusions can only be regarded as posing an untested hypothetical set of categories. The vagueness of the working hypothesis of the study may account for the failure to order the data collected so as to test the conclusions. But if the hypothesis were actually more clear than is stated in the report, then this misalignment between analysis and conclusions must be accounted for by slavish adherence to a set pattern for generalizing statistical data—a pattern that was not modified by the nature of the beginning hypothesis or by the discoveries in the process of inquiry.

It is necessary to make only brief reference to a Michigan study, published in 1941, which bears the results of a study of changes in land use in four northern counties of that state over the period 1925–1940.[31] In this report there is no indication whatever of a hypothesis. The purpose of the study is given simply: "to determine trends in land utilization in this region, with the reasons therefore."

In a review of this work, it was pointed out that the reader "will feel that the large bulk of time, effort, and money that went into the study was spent on the careful collection and organization of data which . . . contribute least to the importance of the work . . . A probable explanation of this lack of balance would seem to be that the study of changes in the area was conceived and set in motion under a given prevailing appraisal of the regional economy; whereas it appears that the changes taking place prove to be of such a type as to require a revision of that very appraisal."[32] The review goes on to point out that certain data in the report are suggestive of the need for an overhauling of the formulation of the cutover problem and a reconsideration of the standard proposals for patterns of action,

[31] H. J. Andrews and W. J. Bromley, *Trends in Land Use in Northern Michigan* (The Charles Lathrop Pack Forestry Foundation, 1941).

[32] L. A. Salter, Jr., "Transition in the Northern Lake States," *Journal of Land and Public Utility Economics*, vol. XVIII, 1 (Feb. 1942).

but that the report does not make use of these suggestive items of information. Instead, it proceeds with a summarizing of quantitative data (none of them direct evidence of experiences of the people in the area) and adheres to an old pattern of explanation and recommendation. As a result some of the most fundamental phases appear to remain untouched.

Other Cutover Areas

Problems associated with the shifting of an area used for lumbering activities to other land uses have not been confined to the upper Lakes States. They have also caused concern in the western and southern forest regions of the United States and have attracted some investigational attention of land economists.

Since the lumber industry moved from the Lakes States to the South and then to the Northwest, the problems which began to be felt in the Lakes States as early as 1912 were not pressing in the other areas until a later date. Also since Minnesota and Wisconsin were two of the very first states to undertake research work in agricultural or farm economics, it might be expected that research attention to the problems of reconverting cutover areas would be greater and earlier in those states.

THE PACIFIC COAST REGION

The sequence of research interest in the South and in the West follows that of the Lakes States region. Although a "cost of land clearing" study in western Washington was published as early as 1912,[33] it was not until 1924 that a farm management study of farms in the western cutover area appeared.[34]

It was not until the depths of the depression of the 1930's that settlement on cutover lands became so widespread, and research personnel became sufficiently available, that land economic work was undertaken. Beginning in 1936 a series of mimeographed reports was issued from the Pacific Northwest region.[35] These reports are more

[33] H. Thompson, *Costs and Methods of Clearing Land in Western Washington* (U.S.D.A. Bureau of Plant Industry Bull. 239, 1912). Also see H. D. Scudder, *Stump Land Reclamation in Oregon* (Oregon Agr. Expt. Sta. Bull. 195, 1922).

[34] E. R. Johnson and E. D. Strait, *Farming the Logged-Off Uplands in Western Washington* (U.S.D.A. Bull. 1236, 1924). Two other reports in the same region on somewhat similar topics are not economic research in any formal sense: W. T. Clarke, *Agriculture in Cut-Over Redwood Lands* (California Agr. Expt. Sta. Bull. 350, 1922); and G. R. McDoles and J. H. Christ, *Farming Practices for Cut-Over Lands of Northern Idaho* (Idaho Agr. Expt. Sta. Bull. 136, 1925). A comparison of these bulletins with those of the Lakes States at approximately the same time emphasizes the relatively far-advanced stage of land economic research at Minnesota and Wisconsin. Also see J. H. Christ, *The Cut-Over Lands of Northern Idaho* (Idaho Agr. Expt. Sta. Bull. 169, 1930).

[35] E. F. Landerholm, R. E. Bell, and A. E. Orr, *Reconnaissance Land Use Classification of Snohomish County, Washington* (U.S. Resettlement Administration, 1936); E. F.

nearly comparable to the Wisconsin Extension series of 1929–1933 (see footnote 16) than to research undertakings. The bulk of attention in the work is given to a delineation of areas that are or are not suitable for farming, suitability being determined by physical soil and topographic features and location. There is no testing of a hypothesis; the nature of the problem and the nature of the solution of the problem are stated but without testing evidence. In other words, these reports are premised on two assumptions: (1) If new settlers are directed away from areas classified as unsuitable and if settlers in areas classed as suitable are assisted in clearing larger acreages, then "a sound economy based on agricultural and forest resources can be established"; and (2) if the land is classified according to its physical capabilities and location, then action to effectuate the first assumption will be expedited. These "if-then" propositions are assumptions of the work; they are not used as hypotheses to be tested and revised.

Finally, as the depression continued and as farm people poured out of the dry Great Plains region to seek some source of economic opportunity on the Pacific Coast, there again appeared an attempt to direct that movement to a desirable outcome. With land economists, sociologists, and farm management specialists cooperating, the U.S. Department of Agriculture launched a series of studies in that region. Some of these investigations dealt with settlement of irrigated areas; but several were concerned with problems of cutover areas. True to form, one of the first of these was a study of the costs of clearing land.[36] Among the other reports published were studies of settlement problems on northern Idaho cutover land[37] and on cutover lands in western Washington.[38]

The western Washington report may be noted briefly in respect to methodological procedures. The study was made to "determine the

Landerholm and R. E. Bell, *Reconnaissance Land Use Classification of Gray's Harbor County, Washington* (U.S.R.A., 1936); R. E. Bell, C. Tjerandsen, and C. H. Dwyer, *A Reconnaissance Land Classification of Boundary County, Idaho* (U.S.R.A., 1936); J. C. Moore, *A Land Use Study of Curry County, Oregon* (U.S.R.A., 1936); J. Tjerandsen, *A Land Use Classification of Latah County, Idaho* (U.S.R.A., 1937); R. E. Reynolds, *A Land Use Study of Tillamook County, Oregon* (U.S.R.A., 1937); J. C. Moore, *A Land Use Study of Coos County, Oregon* (U.S.R.A., 1937). C. Tjerandsen, *A Land Use Study of Benewah County, Idaho* (U.S.D.A. Land Economics Report No. 3, 1939). All of the above reports are in mimeographed form only.

[36] W. W. Troxell and H. J. Voth, *Land Clearing with the Bulldozer* (U.S.D.A., Migration and Settlement of the Pacific Coast Report No. 7, 1941).

[37] *Cut-Over Land of Northern Idaho* (U.S.D.A., Migration and Settlement on the Pacific Coast Report No. 5, 1941).

[38] C. P. Heisig, *Settlement Experience and Opportunities on Cut-Over Lands of Western Washington* (Washington Agr. Expt. Sta. Bull. 399, 1941). The same report is otherwise published as C. P. Heisig, *Cut-Over Land in Western Washington* (U.S.D.A., Migration and Settlement on the Pacific Coast Report No. 6, 1941).

problems encountered by settlers on cut-over lands . . ., how settlers are meeting these problems; what success . . . [they] are having . . ., and what policies and action might be adopted in connection with such settlement." Clearly, this is a problematic situation in everyday experience and the purpose is to find remedial actions to be instituted. If something in the way of new action can be instituted, these settlers will be better able to provide themselves "with a living." The study is based on over one thousand brief interview schedules from all families in five areas and 267 detailed farm schedules from a one-fourth sample of this population taken in 1939.

After pointing out that 60 per cent of the families had moved to their present places in the depression decade 1929–1939 and that half of these people came from nonfarm pursuits, the records are grouped for analysis on the basis of whether or not the farm had been developed before 1929 (irrespective of the settlement of the present occupants) and on the size of the farm enterprise (three groups based on farm man work units). Following comparisons of the averages of farm business items among these six groups, it is pointed out that three of the groups (two new-farm groups and one old-farm group) average less than zero net cash farm returns, that two other groups (one old- and one new-farm group) include no places with relief income, and that all five of these groups average about the same total net cash available for family living. In other words, the analysis most clearly suggests that the basis of sorting the schedules is not significant, but that other factors, treated as minor, probably would have been much more important to an analysis of the experiences, ends-in-view, and needs of the people in the area.

In a section on the financial progress of the families and their clearing of land, different breakdowns of the records are made; but they are based mainly on new and old farms and length of occupancy. Still, from time to time, the report mentions the fact that most of the areas studied are near urban districts, and that nearly a third of the depression occupants have done no clearing at all of land. Yet the conclusions of the report are that public assistance for land clearing is "necessary if most settlers are to be expected to provide themselves with sufficient cleared land for a minimum adequate farm unit." Conclusions along this line are, to be sure, safeguarded in the text by such phrases as, assuming "the operators . . . are capable and desirous of increasing the effective size of their farms," or "on the assumption that a considerable proportion of the settlers . . . are willing and able to undertake the operation of a minimum full-time dairy farm," and so on. These phrases are defensive safeguards to the report, but can it be said that the recommended policies are warranted by the evidence developed in the inquiry?

Once again the limitation on research appears to be owing to a failure (1) to have a developed hypothesis and to use it in structuring the pattern of analysis, (2) to have the hypothesis set in terms of actions instituted and consequences experienced, and (3) to rearrange and rebuild the pattern of analysis when a modification of the hypothesis is suggested by the data unearthed under the hypothesis in its original form.

THE SOUTHERN REGION

As early as 1917, a dawn of a new constructive era was seen by an impressive Cut-Over Land Conference of the South;[39] but, in spite of all the land economics research work on cutover problem areas in other parts of the country, it was not until twenty years later that any important research was reported on economic problems of a southern cutover area, and all of this work was done in the single state of Louisiana.

The first cutover study in the South, like Wisconsin Research Bulletins 134 and 139, was undertaken by sociologists for the purpose of investigating the "human side" of governmental land purchase and resettlement proposals. Suffice it to say in respect to this Louisiana Bulletin 268,[40] which has as its purpose only the "careful inventory" of "basic facts" pertaining to the situation, that it has no hypothesis expressed or implied and no conclusions to indicate in any way how the problematic situation might be resolved. And this description fits a bulletin that deals with a problematic situation real enough to be described as found in an area that is a "severe liability," and engendered by "programs of action being pushed by the Federal Government which are likely" to raise "violent disagreements" and "charges and counter-charges."

A second Louisiana bulletin on a problem situation on cutovers involves an element quite different from most.[41] The company-owned cutover lands involved have been used so long as free ranges by local farmers that their farm enterprises are based on these rent-free grazing areas. The problem situation arises out of the conflict between these farmers and new "patch" farmers who have begun to buy small cut-

[39] *The Dawn of a New Constructive Era,* Cut-Over Land Conference of the South (New Orleans, 1917).

[40] T. L. Smith and M. R. Fry, *The Population of a Selected "Cut-Over" Area in Louisiana* (Louisiana Agr. Expt. Sta. Bull. 268, 1936).

[41] B. M. Gile, *Economic Utilization of Rural Land Resources in Beauregard Parish, Louisiana* (Louisiana Agr. Expt. Sta. Bull. 322, 1940).

over farms from the lumber companies. Not only does the presence of farms of the latter type interfere with the grazing areas of the former farmers, but it "produces contention" over damages to crops and may lead to a "fencing law, which would mean the discontinuance of legal free range for livestock."

The primary data on which the analysis is built, aside from census and *Crops and Markets* reports, are from a brief farm schedule obtained from one hundred and ninety-two farmers. It is clear from the report that no data were collected in respect to the problems of those whose settlement was creating consternation among the established farmers. In fact the only material in respect to the settlers' problems was obtained when "experienced local farmers were asked their judgment" as to what the settlers' needs were. Furthermore, although the census figures show that 73 per cent of the farms of the area are under fifty acres, the two lowest acreage classes in the survey average thirty-six and fifty-five acres respectively and still only account for 46 per cent of the surveyed farms. Finally, for a social inquiry premised on the need to study the "general economic and human problems" of the area and not just "the internal management of farms," it is surprising to find that no mention is made of the distribution of whites and Negroes among either the experienced farmers or the "patch" farmers.

The need to consider color as a strategic factor in land settlement studies in Louisiana is suggested in another Louisiana cutover settlement area study [42] and is a central issue in a later bulletin.[43] This last study is a piece of research work sufficiently impressive to invite attention to some of its characteristics.

Louisiana Bulletin 335 reports on an inquiry into the problems of settlement of a cutover section of the Mississippi Delta. It is not clear from the publication whether the beginning hypothesis was refined to the point of suggesting, for example, that the practices and charges of the land selling agencies and the restrictions of the federal Agricultural Adjustment Administration program were strategic to the problems of the area, or whether these were uncovered during the preliminary work. But it is known that a great deal of factual material was accumulated and made available on a limited basis some time previous to the issuance of the bulletin. That some leading concepts were developed in that early work is suggested by the structure of

[42] Troy Mullins and B. M. Gile, *Economic and Physical Factors Affecting New-Ground Farmers in Madison Parish, Louisiana* (Louisiana Agr. Expt. Sta. Mimeo. No. 12, 1940).

[43] P. E. Jones, J. E. Mason, and J. T. Elvove, *New Settlement Problems in the Northeastern Louisiana Delta* (Louisiana Agr. Expt. Sta. Bull. 335, 1942).

the bulletin which, instead of being set by an item-by-item analysis of field schedules, is based on a series of problems. Although there is some rigidity in the tabulations of some of the data on a standard form of averages for white and colored groups of settlers, the analysis is not bound to it. Qualitative summaries of contract provisions or governmental regulations and conceptual analysis of their implications are woven into informal statistical comparisons, references to other sources of information, illustrative cases, and the more routine tabular presentations. Many parts of these analyses specifically raise this question: Why are these things so? Suggestions are made as to possible remedies for the subproblems that are uncovered.

In short, the procedure used in the study defies a brand name, except that the report shows organization of relevant data and evidence around a progressive inquiry into the problems posed. Only in a section on "social aspects" does there appear to be a discussion of data that adds practically nothing to the analysis, but this is short and without statistical elaboration. Another short section indicates that matters of local public finance were looked into (recalling that at this time, most land utilization studies assumed these to be a critical element), but the report finds "no serious problems at present" and leaves the subject after referring to possible sources of future problems of local government. In a final section, the points made in the subproblem analyses are brought together and a series of recommended remedial actions are advanced with considerable specificity.

In appraising the methodological lessons of this report, it should be borne in mind that (1) a good deal of preliminary searching and probing was engaged in which would not be judged from the printed report alone, (2) the bulletin gives the impression of a high degree of homogeneity within the two settler groups that are recognized, and to a considerable extent it is this sense of homogeneity that lends confidence to the interpretations of the evidence although no formal indication of this uniformity is actually presented.

Recapitulation

HISTORICAL

The first studies clearly reveal the wide differences between the farm management group and the economists. The studies by Boss and other farm management men stress formal and precise presentation of quantitative data, analytical materials that might be called private facts, and a framework of individual action. The study by Ely and Hibbard, on the other hand, shows obvious disregard for quantitative precision, reliance on conceptual reasoning, analytical materials that include group and public facts, and a framework of public or group action. It is in the work of John D. Black, a Wisconsin graduate in agricultural economics at Minnesota, and L. C. Gray, a Wisconsin graduate in land economics in the U.S.D.A., that these two schools of thought are amalgamated.

Later work by Hibbard shows continued wariness of the use of field records from people on the land and continued emphasis on the role of public action in the political economy tradition. Also in his work is the development of new geographic analysis techniques which may be regarded as a speeding up of the localized mapping, which in Michigan has bogged down in physical science precision, and as a slowing down or localization of the geographic method, which to Taylor and Baker meant a very extensive type of map analysis.

Next is noted the work of O. B. Jesness and associates, in which a variety of procedures are joined together and refined. This coalescence, again done at Minnesota, included the use of synthetic projection which grew out of the new farm management budgeting technique and its application to local government by G. S. Wehrwein at Wisconsin.

The development of Lakes States cutover studies also clearly mirrors the influence of the public action programs of the 1930's. Beginning with the series of Wisconsin Extension publications, running through Jesness' work at Minnesota, and included in the work of the U.S. Department of Agriculture is evidence of the redirection of research to the provision of local information necessary to the application of a previously determined method of solution.

Finally, with almost complete discontinuance of direct field investigations, with a centering of attention on local government data and criteria, and with the freezing of concepts around the application of established action programs, the sociologists suggested that the whole structure of inquiry may have been pulled too far away from the ongoing experiences of people.

In general, the development of land utilization research in the Lakes States, particularly in Wisconsin and Minnesota, set the pace for similar work done in other cutover problem areas.

METHODOLOGICAL

This review also gives substance to the analysis of scientific inquiry presented in Chapter III and shows the consonance of the concepts in that chapter and the evidential materials in the chapter.

First, it is clear that these inquiries have been occasioned by the experiences of people; they have arisen out of felt difficulties, confusions, and doubts in human experience. In nearly every case the questions center about what to do to get some postulated result.

Second, it is noted that for most studies there is a formulation of the problematic situation — a definition in terms of possible strategic factors and ends-in-view. But in some cases the problems are only vaguely formulated, and the outcome of the research cannot be regarded as warranted assertions on action to be taken to solve the problem. The assertions can be considered merely hypotheses for further inquiry.

Also, cases are noted in which the problem formulation is not looked upon as flexible but has come to be so fixed as to deflect research from new discoveries of large significance. In the history of land utilization research in the Lakes States cutover, this error is of outstanding importance.

Third, this review indicates the importance of attention to both the statement and the use of hypotheses in directing inquiry. When there is no statement of a hypothesis or only a vague one, irrelevant data are introduced, analyses are made that do not provide helpful evidence, and conclusions are weak, unconvincing, or useless. Where the research shows that there is an operating hypothesis, even though unstated in the publication, the reconstruction of the hypothesis in "if-then" form serves to clarify and thus to help appraise the validity of the conclusions, in terms of the weight and relevance of the evidence presented.

Failure to regard hypotheses as flexible, changing guides is made evident in stereotyped presentations of data. Even where the set format is designed to check some sort of hypothesis, there are instances in which certain conclusions are given that are neither within the frame of reference of the hypothesis nor supported by evidence; yet there may be clues within the data suggesting that were these data rearranged (as a revision of the hypothesis might require), they would be of considerable value in reappraising the problem or in clarifying the conclusions.

Fourth, most of the studies use data or means of preparing data that limit them to conclusions that are suggested but not tested by the evidence. In the work reviewed, 'an outstanding example is the use of tax-delinquent status of land as evidence of one actual process of experience, when, in fact, it might equally well be indicative of any number of complex experiences. Another common example is the use of outside labor receipts as evidence of the farm's being so poor as to force the operator into nonfarm occupations; or the interpretation of residence in open country as evidence of farming, or of decreased timber land assessments as evidence of actual logging operations. Such data may be suggestive indexes of what may have actually happened, but lacking some other carefully integrated data, they are not evidence of human experience. Too often the supplement is provided, not by experiential information but by conjectural history or by frozen concepts of problems.

Also, the methods of handling data usually give suggestive but not testing force to the analysis. Map correlations, for example, may show tax delinquency and poor soils in geographic proximity. These data suggest that farm failures on poor soil result in tax delinquency, but they do not test it. So too with tabular comparisons of cross-classified group averages. Higher average profits for groups with higher income per cow, for instance, may suggest that better grades of cows lead to higher profits; but they do not test it by reporting experiences in which better grades of cows were introduced under stated circumstances and higher profits were the outcome.

Fifth, only rarely can instances be found in which patterns of actual experience are preserved as experimental processes. The testing value of such evidence is noted in those instances where the procedures used make it available. In only one instance are the data specifically given in such a way as to make this type of analysis possible, but the reader is left to work it out for himself. In other instances, such analysis is done almost haphazardly. What there is of it proves to be the most substantial evidence presented for the conclusions reached, but the inconsistent nature of the presentation reduces its effectiveness.

Sixth, and finally, there is only scant recognition of the possibility of seizing upon controlled institution of actions as experiments to test ideas. In the best example of this procedure, a foundation is laid for later completion of the experimental inquiry. In other cases, the opportunity is completely lost under the greater drive to promote use of ideas previously codified in a program of action.

CHAPTER VI

Land Utilization Research in Other Areas

The Old Highlands

As was pointed out in Chapter II, the shrinking of the agricultural margin after World War I not only affected the newly settled areas of the cutover and the arid regions but also accentuated the long-time withdrawal of agriculture from the areas of poor land in the highlands of the East. This contraction was further accentuated in many parts of the East by the prevailing prosperity in nearby urban districts. Under these circumstances there were noticeable changes in the pattern of land utilization and mainly a shift from use of the land for farming to no active use at all—farm abandonment. The decline in rural populations caused concern and soon land economists were called upon to study the situation. The subregions in which work was conducted were primarily central New York State, northern New England, the Ohio Valley highlands, the Appalachian highlands, and the Ozark highlands.

NEW YORK STATE

Historically, attention to problems associated with the contraction of agricultural use of the land can be traced to comments by G. F. Warren in his earliest farm survey of four townships of Tompkins County, New York. He pointed out that farmers should give greater attention to the possibility of profits from practicing forestry on the fields that were poorest for crops or pasture.[1] Although Warren noted many abandoned farm houses, he observed that this did not indicate farm abandonment since "Nearly all the land is rented by neighboring farmers for hay or pasture," and the phenomenon was only a result of the trend to mechanization and larger farm sizes.

By 1923, however, the Chenango, New York, County Farm Bureau was concerned over the existence of idle farms, and it asked for a survey "to determine what types of farm bureau work could best be carried on and should be carried on for the benefit of the community." In response to this request, a preliminary, unpublished survey was made of the town of Pharsalia in the summer of 1923. The survey

[1] G. F. Warren and K. C. Livermore, *An Agricultural Survey* (Cornell Agr. Expt. Sta. Bull. 295, 1911).

concluded that the farms on unproductive soils could not compete for labor with farms in better situations and with city industries, that this situation would continue, and that "forestry, rather than agriculture would probably be desirable as a means of using the idle land." In the following year a more intensive investigation was undertaken in Pharsalia and in it "Parts of the town were recognized as less productive for agriculture than was the rest . . . and they were recommended as suitable for forestry projects." [2]

This statement of the purposes and origin of the Pharsalia survey is not given in the report on Pharsalia. In the published bulletin the only specifically stated object is "to present as complete a picture as possible of the regional conditions." [3] At the end of the Pharsalia report, however, is a section entitled "Possible Forest Areas in Pharsalia," and there data are given to indicate the probable cost of purchasing these areas for public forests. Insofar, then, as this investigation was designed to determine which areas should be put into public forestry, the prior determination that a public purchase and reforestation program was the solution to the problematic situation posed by the Chenango Farm Bureau was based on the preliminary, unpublished survey conducted in the summer of 1923.

During the same year the Pharsalia report was printed, another bulletin was issued in which an almost identical presentation of quantitative data for 13 additional areas in the state is given.[4] In this report it is stated that "the real problem was to locate definitely the areas of abandoned farms and to obtain all the information possible concerning them, for it would then be possible to make intelligent suggestions as to how the land might be brought into some profitable use." The bulletin may be characterized as a 285 page compendium of the most detailed sort of quantitative facts pertaining to these 13 areas. But like the supplementary report on Ellery in Bulletin 476, the studies of these 13 towns do not include the separation of possible forest areas within the larger areas surveyed.

The summary of all 15 areas shows that these areas contain 685 occupied farms (one fourth of which were used not as farms but only

[2] This history is given in A. B. Lewis, *Methods Used in An Economic Study of Land Utilization in Tompkins County, New York and in Other Similar Studies in New York* (Cornell Agr. Expt. Sta. Mem. 160, 1934).

[3] William Allen, *The Utilization of Marginal Lands* (Cornell Agr. Expt. Sta. Bull. 476, 1929). In this printed bulletin the report on Pharsalia is followed by nearly identical summary of data for Ellery town in Chautauqua County, except that no forest purchase areas are delineated because "The valuations placed on much of the land of this area are too high for a forest project." But "It would be advisable to convert some of the most rugged and inaccessible parts of the farms to forest."

[4] L. M. Vaughan, *Abandoned Farm Areas in New York* (Cornell Agr. Expt. Sta. Bull. 490, 1929).

as homes), 764 vacant farms, and 329 odd parcels of land. The conclusion of the report is that for "the vast area of idle land," in order to keep it "from agricultural production and still not remain wholly idle and unproductive, forestry has been suggested as the logical remedy. . . . A definite reforestation program seems desirable for the cleared land. . . . Apparently the State must do most of the work if it is to be done."

As noted in Chapter II, in 1929 the State of New York authorized purchases of areas of not less than 500 acres for reforestation, in 1930 it began to appropriate funds for a survey of land utilization and land classification in the state, and in 1931 it adopted a plan for acquiring and reforesting a million acres of land. Under this program, the first county survey of land classification and utilization—one for Tompkins County—was published in 1934.[5] Since that date very similar reports and maps have been published for 15 additional counties.[6]

The New York procedure for marking out "possible forest purchase areas" has attracted a good deal of attention and, as will be pointed out below, has been more or less duplicated in some other states.[7]

Beginning with the first Tompkins County work, the only specifically defined classes of land have been (I) land "primarily adapted to forest and recreational uses," and (II) land "better suited to forestry and recreational uses than to farming." The need for such delineations to aid in the administration of the state reforestation program is clear. Beyond this, however, the New York work has classified other rural lands into as many as six higher categories, all of which cover "agricultural lands" of increasing "intensity of use." These higher classes have been designated as priority areas for the extension of rural road and electrification improvements and extensions. Also, because a credit study showed that the foreclosure rate

[5] A. B. Lewis, *An Economic Study of Land Utilization in Tompkins County, New York* (Cornell Agr. Expt. Sta. Bull. 590, 1934).

[6] These bulletins, all with the same title as that above except for the county name, and all by the Cornell Agricultural Experiment Station are: F. F. Hill and G. T. Blanch, *Montgomery County* (Bull. 613, 1934); T. E. LaMont, *Chemung County* (Bull. 640, 1936); T. E. LaMont, *Broome County* (Bull. 642, 1936); P. B. Jones, *Tioga County* (Bull. 648, 1936); H. S. Tyler, *Chenango County* (Bull. 654, 1936); J. M. Efferson, *Genesee County* (Bull. 668, 1937); W. E. Keeper, *Steuben County* (Bull. 674, 1937); R. S. Beck, *Rensselaer County* (Bull. 675, 1937); W. T. Wilson and J. N. Efferson, *Monroe County* (Bull. 683, 1937); E. H. Matzen, *Cortland County* (Bull. 685, 1938); O. H. White, *Clinton County* (Bull. 689, 1938); H. R. Kling, *Wyoming County* (Bull. 707, 1938); Alexander Joss, *Chautauqua County* (Bull. 720, 1938); M. D. Woodin, *Yates County* (Bull. 728, 1940); Alexander Joss, *Otsego County* (Bull. 791, 1943).

[7] See references to the New York work in *Land Classification in the United States* (National Resources Planning Board, 1941); *The Classification of Land* (Missouri Agr. Expt. Sta. Bull. 421, 1940); Charles Gooze, *Progress in Rural Land Classification in the United States* (U.S. Resettlement Administration Land Policy Circular Supplement, Dec. 1935).

was higher on farms of low value and poor land than on better farms, the classification system was suggested as a guide to farm credit leaders.[8]

It is unnecessary here to describe in detail the procedure used in the New York system of land classification. It is enough to say that the classes are designated in all the bulletins by a visual combination of a series of maps showing (1) a rating of the size and condition of farm buildings, (2) land cover, and (3) soil type. Field work emphasized the rating of farm buildings, which can be done relatively quickly, and the more time-consuming recording of land cover by ten-acre squares. Also, as a result of the many farm surveys made over the years at Cornell, some farm management data were usually available for indicating farm returns in the various areas.

That intensity of land use (interpreted as productive man work units required for various forms of land cover) and the size and condition of buildings are indicative of farming returns has been criticized elsewhere, both on theoretical and empirical grounds.[9] But here a different question is raised: How does the land classification work in New York measure up as scientific inquiry, especially since all the reports on it are called economic studies of land utilization?

It is clear from any or all of these reports that the objective of the work in each study was not to investigate, to probe, and to bring a solution of whatever problems in land utilization might be found, but to apply the established classification technique, on the grounds that this information will "assist those who are interested in the land to use it for those purposes for which it is best adapted." But those purposes are, by definition, public purchase and reforestation in land class areas I and II or farming in the higher land class areas.

In other words, in the New York work the formulation of the problem and the hypothesis, the procedures, and the solution are all standardized. The problem from the first was seen as that of encouraging farming on good land and forestry and recreation on poor land. If there is a hypothesis, it is that if poor lands are reforested and good lands are given telephone and electric service, then there will be better use of resources. The procedure is to demarcate these areas in the manner described. The solution is not to test the hypothesis

[8] See Lewis, *Methods Used;* F. F. Hill, *An Analysis of the Loaning Operations of the Federal Land Bank of Springfield* (Cornell Agr. Expt. Sta. Bull. 549, 1932); and *Classification of Land,* pp. 58–74.

[9] P. E. McNall, "Farm Buildings as Evidence of Productivity of Crop Land," *Journal of Land and Public Utility Economics,* vol. XVII, 2 (May 1941); C. H. Hammar, "Intensity and Land Rent," *Journal of Farm Economics,* vol. XX, 4 (Nov. 1938); *Classification of Land,* pp. 75–77, 189.

but to indicate where it applies. The evidence does not consist of materials which test whether the problem is properly formulated or whether the hypothesis is substantiated, but is rather an orderly compilation of facts that only serve the circular purpose of indicating that the system of land classification does, in fact, include, in areas of the higher class, farms with higher average incomes and more substantially developed farm businesses.

Even if it be granted that the problem might have been properly formulated at first for areas such as Pharsalia, and that the early evidence tested the concepts which led to the first applications of the procedure, and even if it be granted that the procedure helps in the administrative task of finding sites that might be considered for reforestation purchase, the procedure is not one that is very useful for the general purpose of exposing and assisting in the solution of existent problems in land utilization.

Evidence on this point is most clear in the case of the Monroe County report. This county, far from being a hilly and relatively isolated section, contains the city of Rochester with a third of a million people. Yet even here the same procedures of studying land utilization are applied as were applied in the remote hilly areas of south central New York. On the very first page it is stated that in Monroe County all farms should be served by hard roads and electric lines; and in the early section on definitions of land classes, it is stated that "Land classes I and II were not mapped in Monroe county."

In view of the fact that these studies all define only land classes I and II in terms of the need for adjustment of land utilization, the conclusion obviously is that no land in the county is better suited to forestry or recreational uses than to its present use. But this is a doubtful conclusion for a large county that completely surrounds a city the size of Rochester. The point to be made here is that there was no attempt to probe and expose and solve the emerging problems of land utilization changes in Monroe County. The study may have revealed that there are no evidences of low income farming in Monroe County, but this is evidence relevant only to the reformulation of a problem concept that is set in terms of finding recreational outlets for a huge urban population.

To maintain that the study in question was not designed to serve the purpose of assisting in the solution of the problem of urban impacts on rural lands in such a situation is merely to hold that an "economic study of land utilization" of a county ought not to be expected to assist in the solution of the real problems of the area. Furthermore, the bulletin cannot even be viewed as a preliminary probing of a problematic situation, for there are no references whatever to problems or sources of problems other than the one that years before had appeared to represent the situation in an entirely different geographical setting.

When a problem concept, a hypothesis, and a procedure are applied with great quantitative accuracy in an area where it is obvious from even a cursory look that they are not relevant, and when other prominent strategic factors in the problematic situation are simultaneously disregarded, land economics investigation falls short of meeting the requirements of purposeful inquiry. Yet these very procedures have been defended on the ground that they are scientific because they are quantitatively accurate. But they are scientific only in the Pearsonian view that they result in an orderly resumé of collected facts.

It may also be noted that another possible source of difficulty may be in the failure to structure the observations made in terms of the actual process of experience of the people involved in the problematic situation. That such an analysis might have forced a reconsideration of the problem formulation, hypothesis, procedure, and solution is suggested by some submerged items of data not only in the studies made near large cities, but also in the earlier studies made in the less industrialized parts of the state. Also, later studies in New York made both by land economists and by rural sociologists have thrown these deficiencies into sharp relief.

In the earliest study of Pharsalia, a good deal of confusion exists in the references to vacant farms, farms that are occupied but not operated as farms, and farms that are farmed. Most of the detailed data apply to operated farms; the existence of occupied but unoperated farms is certain but nowhere clearly described. In the later summary of 15 areas, as has been pointed out, it was only noted that a fourth of the occupied places were not operated as farms.

In all the later land classification surveys, the buildings were listed separately as farms or as rural residences. Even in Tompkins County, 25–35 per cent of the buildings in the areas of better land were classified as rural residences rather than farms. Also, the statement is made that rural parts of the county experienced population growth greater than the city of Ithaca. Yet the land classification procedure was not affected by these observations. In Monroe County about 40 per cent of all buildings in each of the land classes were rural residences.

In some studies, for areas with a high proportion of the buildings listed as residences rather than farms, the letter *R* was added to the land class number. Yet in Rensselaer County 23 per cent of the buildings in land class IV were still rural residences as compared with

30 per cent in class IVR. In terms of recommendations the only change to be noted is that land classes IR and IIR are defined as areas that should be purchased by the state if they can be secured "at a reforestation price" but that they "probably will remain in private ownership."

Through all of the work, however, the original and basic concept developed in the earliest work — that the problem was one of encouraging commercial agriculture or reforestation — has persisted. Only recently has a special investigation, aimed at finding out what the people of the area were actually doing, come to the conclusion that in the heart of the old highland areas of south central New York, if industrial activity continues, half the present full-time farms in land classes III–VI will pass over into part-time farming and rural residence and that many places in land classes I and II will continue or will come to be used for the same purposes.[10]

In New York, as in Minnesota and Wisconsin, studies were undertaken to investigate the land purchase and relocation programs. One of these studies merely sets out to help answer this question: "What happens to the persons on land sold to the State for reforestation and recreation?"[11] The report presents tables of facts relative to these people and the places they sold, but the only direct information that is evidence for testing the outcome of the experiment is the answers given by the sellers to the questions asking if they would buy their land back if they could and if they considered themselves better off for having sold. There is no other evidence as to the outcome of the experiment in terms of its specific objectives. One reason why the study is not more enlightening on this point is that these objectives are not clearly stated so that they may be used as a test. Another is that practically all the data are given in a set of standardized tables in which the sellers are grouped according to whether they had lived on and sold their places, lived on and sold a part of their places, or had never lived on their land. Although this classification has in it an important element of sellers' experiences, it is not directly related to the act of state purchase, because two thirds of those who had lived on their land and moved off had done so prior

[10] H. E. Conklin, "The Rural-Urban Economy of the Elmira-Corning Region," *Journal of Land and Public Utility Economics*, vol. XX, 1 (Feb. 1944). Also see F. F. Hill, H. A. Johnson, and D. R. Rush, *Erin* (New York State College of Agriculture, 1943). For a contrast in problems recognized, compare the land utilization survey of Monroe County (cited) with L. B. Tate, *The Rural Homes of City Workers and the Urban-Rural Migration* (Cornell Agr. Expt. Sta. Bull. 595, 1934), which appeared three years before the land classification bulletin.

[11] T. E. LaMont, *State Reforestation in Two New York Counties* (Cornell Agr. Expt. Sta. Bull. 712, 1939).

to the time the state bought their land. Yet all the information is structured on the larger grouping so that the facts relevant to those who actually moved off because of state purchase are hopelessly intermixed with what might have otherwise been a good control group—those who moved out without the state's intervention.[12]

The same type of observation also applies to a second New York study of land sellers, which, although a more comprehensive analysis, still fails to arrange its facts so as to expose the explanations sought.[13] Thus in this bulletin an early section develops and applies to all sellers an index of "cultural type conformity" said to be "useful in understanding" "responses to the government program." Yet no further reference is made to the grouping of families into standard and deviate classes according to this index, until a closing section which merely reports that identical percentages of standard and deviate families made successful adjustments!

Again, an early statement emphasizes the presence in the observed population of some young families with many young children. In a closing section it is said that young families with children made a noticeably better adjustment than other family types; yet no further use is made of this apparently important group of cases in the analysis.

LAND CLASSIFICATION GENERALIZED

During the 1930's there was a general outbreak of interest in land classification work per se. No doubt stimulated by the inventory work in the Lakes States, physical scientists reoriented some of their field survey work, trying to arrive at composite summary map forms that could be generalized as "natural land type," "productivity," or "land form" classification.[14] In these, a chief characteristic is that more emphasis is given to what is on the land; and frequently these

[12] It may be noted that at the end of this report are brief case reports on the situations under which some 22 parcels were transferred. But any similarities or differences among these reports are left to the reader's determination.

[13] Nelson Foote, W. A. Anderson, and W. C. McKain, Jr., *Families Displaced in a Federal Sub-Marginal Land Purchase Program* (Cornell Agr. Expt. Sta. Rural Sociology Mimeo. Bull. 11, 1944).

[14] Examples are C. E. Walker, *Land Survey of the Town of Durham, New Hampshire* (New Hampshire Agr. Expt. Sta. Bull. 255, 1931); J. O. Veatch, *Agricultural Land Classification and Land Types of Michigan* (Michigan Agr. Expt. Sta. Special Bull. 231, 1933); G. D. Hudson, *The Rural Land Classification Program* (U.S. TVA, 1935); F. B. Howe, *Classification and Agricultural Value of New York Soils* (Cornell Agr. Expt. Sta. Bull. 619, 1935); B. E. Gilbert, *A Study of Land Utilization in Washington and Kent Counties, Rhode Island* (Rhode Island Agr. Expt. Sta. Bull. 261, 1937); G. G. Pohlman, *Land Classification in West Virginia Based on Use and Agricultural Value* (West Virginia Agr. Expt. Sta. Bull. 284, 1937); M. F. Morgan, *The Social Characteristics of Connecticut Land Types* (Connecticut Agr. Expt. Sta. Bull. 423, 1939); R. E. Storie, *Natural Land Divisions of Santa Cruz County, California* (California Agr. Expt. Sta. Bull. 638, 1940); L. A. Wolfanger, *Land Farm Types* (Michigan Agr. Expt. Sta. Tech. Bull. 175, 1941); A. B. Beaumont, *Natural Land Types of Massachusetts and Their Use* (Massachusetts Agr. Expt. Sta. Bull. 385, 1941).

land cover maps are termed land use or land utilization maps. As a result, there is often some confusion as to whether a land use or land utilization classification is based only on physical features or whether it recognizes the economic operating unit within which the land is used.

Like the general soils survey, which had proceeded county by county in the United States for many years, most of these new physical surveys were done with the broad purpose of making information available for whatever uses individuals might wish to make of it. However, one outstanding soils scientist has correctly argued, in presenting a technique in which he combined physical with economic location features, that "To arrive at an effective, practical and accurate method, the use or objective of the classification must be clearly borne in mind."[15] And later the U.S. Soil Conservation Service, with its own program needs in mind, developed a system of physical land classification based on erosion susceptibility.[16] But, since the sponsoring agency is nationwide, there may be some tendency for its classification system to take on a general purpose connotation under the title of a "use capability" classification of land.

Land economists gave some attention to special types of land classification for tax assessment and credit appraisal purposes,[17] but they too attempted to develop general purpose classification systems.[18]

[15] C. E. Kellogg and J. K. Ableiter, *A Method of Rural Land Classification* (U.S.D.A. Tech. Bull. 469, 1935).

[16] As an example: J. G. Steele and R. G. Mowrey, *Erosion and Related Land Use Conditions on the Crooked Creek Project Near Indiana, Pennsylvania* (U.S.D.A. Erosion Survey 16, 1940). Also see E. A. Norton, *Soil Conservation Survey Handbook* (U.S.D.A. Misc. Pub. 352, 1939).

[17] A. J. Englehorn, *Land Classification as a Basis for Land Appraisal and Equalization of Tax Assessment* (Story County, Iowa) (U.S.R.A. Land-Use Planning Pub. 8, 1936); C. H. Hammar, "An Approach to the Grading of Land for Purposes of Appraisal," *Journal of Farm Economics*, vol. XVIII, 3 (Aug. 1936); *Classification of Land, passim.* For a time the U.S. Farm Credit Administration tried the New York system of classification as an administrative device for improving their loan determinations. See O. H. White, *The Productivity of Land in Relation to Farming Returns, Farm Capital, and Loan-Paying Capacity* (Frederick County, Maryland, 1938) (U.S. Farm Credit Administration, 1941).

[18] R. E. Willard and N. W. Johnson, *Present Land Uses* (Washington) (Washington Agr. Expt. Sta. Bull. 288, 1933); P. I. Wrigley, *Land Use in Pennsylvania* (Pennsylvania Agr. Expt. Sta. Bull. 317, 1935); M. H. Saunderson *et al., An Approach to Area Land Use Planning* (U.S.R.A. Land-Use Planning Pub. 16, 1937); C. H. Hammar and H. Jenny, "Land Classification as a Basis for Land Use Planning," *Journal of Farm Economics*, vol. XVI, 3 (July 1934); C. P. Barnes, *Land Classification: Objectives and Requirements* (U.S.R.A. Land-Use Planning Pub. 1, 1936); P. E. Eke, *Basic Data for Land Classification* (Idaho Agr. Expt. Sta. Bull. 232, 1939); R. W. Harrison and P. L. Searfose, *Classification of Agricultural Areas, Frederick County, Maryland* (Maryland Agr. Expt. Sta. Bull. 440, 1941); David Weeks and H. R. Josephson, "Economic Criteria for Classifying Non-Urban Land According to Probable Best Use," *Journal of Farm Economics*, vol. XXI, 2 (May 1939). An unusual contribution has been made in a late study which has a general purpose objective yet recognizes the need for discussing land margins in terms of a specific use. H. C. Woodworth and J. C. Holmes, *Dairy Opportunity Areas in New Hampshire* (New Hampshire Agr. Expt. Sta. Bull. 340, 1942).

Many states took up the New York land classification procedure. Since its justification had grown from selecting areas suitable to public purchase for reforestation, to designating priorities for road improvement and electric and telephone line extension, and to providing a basis for credit-risk determinations, it was easy to extend the list of possible uses. Essentially the same land classification technique came to be used for the purpose of "presenting information which will aid in establishing a better relationship between people and the land." In addition to those already mentioned, possible specific uses were listed: determining school districts, fixing fire insurance rates, aiding in farm management teaching and research, helping buyers and sellers of land, assisting in the administration of governmental programs of all sorts, and laying a basis for rural zoning.[19]

The New York classification system, which had inherent recommendations for action in the areas classified, was used not only on the grounds that it was helpful to many widely different purposes. It was also applied to the big potato areas of the remote corners of northern Maine, to the highly urbanized area around Wilmington, Delaware, and to the hill towns of Vermont, just as within New York State it was laid blanketlike over the poor hill sections of Tioga County and the metropolitan region of Troy. And it was based on a formulation of land utilization problems as a choice between commercial farming and the purchase of land at four dollars an acre for reforestation purposes.

THE YEARS 1930–1933

The early land utilization work in eastern United States consisted of analyses of soil types and of the uses of fields within farm units.[20]

[19] See R. O. Bausman, *An Economic Study of Land Utilization in Kent County, Delaware* (Delaware Agr. Expt. Sta. Bull. 224, 1940); R. O. Bausman, *An Economic Study of Land Utilization in New Castle County, Delaware* (Delaware Agr. Expt. Sta. Bull. 229, 1941); R. O. Bausman, *An Economic Study of Land Utilization in Sussex County, Delaware* (Delaware Agr. Expt. Sta. Bull. 233, 1941); W. E. Keeper, *An Economic Study of Land Classes, Blair County, Pennsylvania* (Pennsylvania Agr. Expt. Sta. Bull. 413, 1941); A. E. Watson, *A Study of Land Use in Thirty-One Towns in Aroostook County, Maine* (Maine Agr. Expt. Sta. Bull. 413, 1942); A. E. Watson, *Land Classification in Waldo County, Maine* (Maine Agr. Expt. Sta. Bull. 417, 1943); W. L. Gibson, Jr., *An Economic Land Classification of Halifax County* (Virginia Agr. Expt. Sta. Tech. Bull. 86, 1943). Also see J. Hitchcock, "An Agricultural Classification of Vermont Lands," *Vermont Farm Business*, vols. 4, 6, 1937); C. E. Allred *et al.*, *Economic and Social Factors Associated with Land Class in Jefferson County, Tennessee* (Tennessee Agr. Expt. Sta. Mimeo. 76, 1938); C. E. Allred and A. C. Robison, *Farm Organization by Land Class on the Eastern Highland Rim* (Tennessee Agr. Expt. Sta. Mimeo. 89, 1939); A. W. Peterson, M. T. Buchanan, and B. D. Parrish, *Economic Land Classification in King and Snohomish Counties, Washington, and Its Influence on Full-Time Farm Returns* (Washington Agr. Expt. Sta., 1944).

[20] For examples see W. I. Myers, *An Economic Study of Farm Lay-Out* (Cornell Agr. Expt. Sta. Mem. 34, 1920); I. G. Davis and C. I. Hendrickson, *Soil Type as a Factor in Farm Economy* (Storrs Agr. Expt. Sta. Bull. 139, 1926).

When late in the 1920's the Division of Land Economics helped to launch a series of land utilization studies in the highlands, this emphasis persisted in the first report published.[21]

The Pennsylvania study illustrates the difficulty involved in reorienting research procedures to a new form of problem. While the problematic situation is described in terms of the "tendency towards farm abandonment," the purpose is only "to determine what factors make land submarginal for farming." And then, the heart of the procedure is an intensive study of a sample of 50 farms, specifically selected as above average in income and stable in occupancy. Further, although a reconnaissance showed that on only 68 per cent of the census farms were the families principally engaged in farming, the selected sample includes only a few on which outside labor was important.

The farm management analysis in the bulletin is not a routine one because the items are presented for each of the 50 farms surveyed, and, in addition, a statement is given as to the "reason for size of labor income" on each of the fifty farms. This is followed by a brief farm management case report for 14 various farm situations. Yet nowhere in all this farm-by-farm information is there a bringing together of apparently strategic factors bearing on abandonment processes among these farms.

Following the farm management material, a brief statement on forestry leads immediately to the presentation of a map on which are shown the four areas in the county "recommended for forestry and recreational purposes." The meager information given about these areas is not much enhanced by the solitary farm (among the sample of 50) that is in the forestry areas. A few sentences say that these areas are isolated, are of high altitude, and have but few inhabitants and practically no farming. The conclusions of the study consist merely in the prediction that the recommended forestry areas will be abandoned, that there will be "no far-reaching changes" in one part of the county, that poor soils will be farmed less intensively, and that the county will not experience general abandonment. In sum, the study was not aimed at testing a hypothesis involving the search for a problem solution. Similarly the data collected are not such as would test a hypothesis or expose a problem.

Two highland studies conducted in southeastern Ohio and published early in 1930 also dealt with areas experiencing farm abandon-

[21] P. I. Wrigley, *Land Utilization and Farm Management in Wyoming County* (Pennsylvania Agr. Expt. Sta. Bull. 257, 1930).

ment and general decadence.[22] The central materials in each of these reports are a series of maps showing land cover, land abandonment, and land assessments by sections, and a series of tables giving data on land cover, condition of buildings, tax valuations, tax delinquencies, and schools by townships. The purpose of the studies is "to throw some light on" the problems of poor land areas and "to serve as a basis for further studies." The conclusions are consistently vague: "The trends taking place . . . should be taken into account by the officials . . . administering the affairs of local government"; and the "use of the idle and abandoned areas may well be considered in any state conservation and forestry program."

The scientific status of these two Ohio reports and the Pennsylvania report is that they are a part of the preliminary phase of looking over some data and noting some prevalent ideas and suggestions in respect to the problematic situation. Their research worth, then, depends on their usefulness in setting the stage for more complete inquiries, and not on the recommendations for action that are given or implied.[23]

One of the earliest large-scale studies in land utilization undertaken by the Division of Land Economics was in Nicholas and Webster counties, West Virginia.[24] Although started in 1926, the report was not issued until 1932, by which time the project had been shunted around until no one apparently wished to assume responsibility for it. The acknowledgments indicate that "In completing the manuscript it became necessary to make extensive changes for which the original authors are not responsible"; and a footnote on page one explains that "it is too much to expect that the results will serve as a model for projects of this type." Although it is, therefore, unnecessary to dwell on the characteristics of this investigation, it may be pointed out that the stated purposes of the study are "to show which type of land should be used for agriculture and which for forestry" and "to indicate lines of improvement in the utilization of both farm and forest lands."

In a classification of land as agricultural or nonagricultural, based on the returns from agricultural enterprise on various soil types, the following facts are noted: (1) Over half the so-called farms on the

[22] J. H. Sitterley, H. R. Moore, and J. I. Falconer, *Land Utilization in a Southeastern Ohio County* (Ohio Agr. Expt. Sta. Bull. 485, 1931); and *Land Utilization in Lawrence County, Ohio* (Ohio Agr. Expt. Sta. Bull. 514, 1932).

[23] G. E. Young, *Marginal Farm Land in Southern Indiana* (Purdue Agr. Expt. Sta. Bull. 376, 1933), is not unlike these Ohio bulletins, except that the analysis is more intensive in contrasting two types of demarcated areas.

[24] Millard Peck, B. Frank, and P. A. Eke, *Economic Utilization of Marginal Lands in Nicholas and Webster Counties, West Virginia* (U.S.D.A. Tech. Bull. 303, 1932).

nonagricultural lands are "mainly country residences"; (2) the percentage of abandoned farms was not much greater on the nonagricultural soils than on the agricultural soils; (3) "a considerable extent of the soil which has been designated here as being suitable for agricultural use is associated with a type of topography unsuited to agriculture. That a beginning problem of formulation and an initial technique of procedure might be found to be in need of alteration may be accepted. But that the concept of the problem and a procedural technique should go so far as to result in these statements in the printed report should be taken as indication of the insufficiency of the analysis that lies behind some research reports, in spite of the fact that they may end, as this one does, with an outline of land utilization recommendations for the survey area.

Another local land utilization survey was made about 1930 in the back hill towns of Vermont.[25] The bulletin is significant because the central feature of the formulated problem, abandonment of farm land, is carefully analyzed as a process of experience up to a certain point and is then discarded in the critical analysis of consequences. The report also illustrates the failure to search, in inquiry, for those elements in the process under study which are most likely to suggest ways and means of resolving the problem.

Briefly, the first third of the bulletin is a good example of bringing a historical review up into the present so that the existing situation is clearly seen as an ongoing historical process. It also shows care in the interpretation of evidence, the comparison of facts from various sources, the use of sample cases, and the cross-checking of static cross-tabulations of data and of dynamic classifications of data arranged by the sequential pattern of experience within the cases.

In spite of this elaborate and impressive analysis of how the abandonment process occurs, when the report approaches the question of the impact of this process upon family incomes, the whole structure of farm classification is shifted from an abandonment basis to classification based on sources of income, size of farm, and the like, and no direct relations are shown between the steps in agricultural decadence and the impoverishment of the farmers. Similarly, the public finance analysis, although it includes figures on the acreages, ownership, and levies on lands in various stages of abandonment, does not bring out any direct interrelation between governmental problems and abandonment.

Had this Vermont study given more attention to the facts that ap-

[25] C. F. Clayton and L. J. Peet, *Land Utilization as a Basis of Rural Economic Organization* (Vermont Agr. Expt. Sta. Bull. 357, 1933).

pear exceptional to the main hypothesis, it is possible not only that the conclusions might have been different but that they might have shown closer integration with the evidence in the analysis. Two examples illustrate this point clearly. First, the very fact that full emphasis is placed on the decadence of these towns should have encouraged a careful analysis of any facts which gave sign of an opposite trend of events; and there is such a sign in a few meager acknowledgments of the existence of a summer home development. But there is no more than a mention of this phenomenon, and no reference to it is made in the concluding program for land use. Today, as is well known, the summer home business has become the backbone of Vermont's land utilization program.[26]

A second example is that although the problem is formulated partly in terms of local government problems created by the draining away of the "vital elements of the economy," the report shows that there is no tax delinquency even by nonresident owners of abandoned land — even though tax rates have increased and even though local public services have been improved. Here again, instead of passing over this countertendency, perhaps one of the most helpful things that this study (or a subsequent one) might have done would be to figure out why the owners of these lands (worthless on the basis of the premises of the inquiry) hold onto and pay taxes on them. But these considerations do not figure in the conclusions even as suggested lines for further inquiry.

Another study of the same period, made in Laurel County, Kentucky, stands as a contrast to the West Virginia bulletin and as an improvement over the Vermont report.[27] The bulletin exemplifies a number of the suggestions made in this thesis. Although the report opens with a general economic history of the area, a pedestrian statement of objectives (to determine present uses, to relate utilization to physical factors, and so on), and a review of census farm data and physical features of the area by districts, this preliminary work leads up to the statement that "This preliminary division of the territory makes possible a more precise definition of the problem." There follows a series of questions as to what apparently needs to be known in order to furnish explanations and to suggest adjustments.

The analysis of physical land utilization facts is so handled, first by fields of various classes and then by operating farms, that the land

[26] Ten years after this study was made and after a sharp growth of summer homes had already taken place, a study was made of the movement. See J. C. Blum, *Land Utilization in New Hampshire Summer Homes and the Rural Economy* (New Hampshire Agr. Expt. Sta. Bull. 344, 1942).

[27] C. F. Clayton and W. D. Nicholls, *Land Utilization in Laurel County, Kentucky* (U.S.D.A. Tech. Bull. 289, 1932).

clearing and cropping sequence is both suggested by cross-sectional data and evidenced by the classification of fields in terms of the sequence in which each field actually was utilized. Then economic data on costs and returns are applied after the practices in experience have been established.

In the analysis of farm organization, it should be noted that the farms are subclassified in various tables, as for example, by location, topographic formation, size, and farm returns. This subclassification reduces the number in each category to relatively small numbers. First, however, the report notes that the "comparison of averages" is not intended to be conclusive but that "clearly defined" patterns that are "consistent with the pattern suggested by collateral data" do provide a basis for qualitative analyses. Also, the number of cases is sufficiently small that in the analysis there are explanations as to why individual or small groups of cases appear as they do in the subclassification comparisons.

Having then amplified "the picture of the cycle of land utilization previously presented, to include the farm unit," the study enters upon a further subclassification in series—location, topographic formation, size and source of income. At this point, however, the original farms are folded into a larger sample, but the bases of the subseries patterns are retained and illustrative case references are reported.

Unfortunately, the later analyses of population characteristics, migration, and mobility do not follow in the same pattern, except in respect to location; but even in these analyses, original devices are used to trace patterns of characteristics and sequential series of changes.

Finally, the conclusions and recommendations given in this report are exceptional as compared with those found in most other studies of decadent areas. The conclusions not only show a sensitivity to the variations in conditions exposed in the body of the report, but they carry a firmness and a positiveness that are distinctive. For example, instead of blanketing the county with a retirement area and an agricultural area, the report recognizes that some sections physically unadapted to agricultural use, be regarded as agricultural in the sense that they will continue to be occupied. The fact that off-farm labor incomes can be as important to rural residents as an increase in their farming income is recognized. Further, more roads and better educational facilities are suggested as a means of aiding population adjustments in some areas.

In sum, this study tackles a problematic situation that is only vaguely defined, arrives at a more specific definition of the problem, and probes it by a careful analysis. The analysis moves from physical land utilization through to the resulting economic and social difficulties. Using a variety of facts processed by various techniques, the study succeeds in exposing the various patterns of experience in the lives and practices of the people. And in the end, the inquiry results in a carefully developed series of suggestions that appear to fit the evidence, rather than a fixed set of routine program recommendations. Although the study does not go so far as it might in all particulars of testing all the recommendations against clear-cut patterns of experienced processes, it nevertheless ranks high in the degree to which it approaches this standard.

THE YEARS 1937–1943

During the years 1934–1936 when action programs for land utilization were being initiated most rapidly, no research results were published in the East except those already mentioned for New York. As was true of the northern Lakes States, studies reported in the latter part of the 1930's in the old highland areas were affected by the existence of federal programs for submarginal land purchase and settler relocation, all of which were premised on a farming or forestry choice of alternatives.

In 1937, a study of "Problems in the back highland areas" of southern Grafton County, New Hampshire,[28] was aimed at a problematic situation defined as one in which "The abandonment of farms . . . has resulted in the isolation of many rural families," stranded people, who have not shared in modern progress.

This problem formulation is not translated into a hypothesis, the objective being only "to isolate and describe the problems" and to lay "a foundation" for "public policy." It is interesting to note that a good deal of the data in the study suggest a different formulation of the problem, and even the conclusions reveal that these new considerations were recognized. Yet nowhere is there a definite reformulation of the problem and consequently the data clearly fail to expose the situation and test the conclusions.

The historical section of the report, which is particularly sketchy for the recent past, emphasizes the abandonment-decadence process in terms of which the original formulation of the problem is drawn.

[28] H. C. Woodworth, M. F. Abell, and J. C. Holmes, *Land Utilization in New Hampshire, I.* (New Hampshire Agr. Expt. Sta. Bull. 298, 1937). The lack of printed research between the pre-New Deal studies and 1937, far from indicating lack of land economics interest, reflects the greater attention in the period to administrative planning and investigation and the lack of attention to exploratory research. See Chaps. II and V.

Under this concept, the item-by-item summary of 252 records of farm business, revealing the existence of limited land resources, small numbers of cows (only a fifth of the farms had over five cows each), small numbers of hens, self-sufficient agriculture, low farm sales, and the like, would seem to support the interpretation that these farms are the end result of a steadily declining agricultural economy.

But later, in an item-by-item summary of the 252 family records, it is noted that nearly half were families "who had moved to the farm in the last decade" and that "many" had "migrated to the farm on account of health, inheritance of property, or desire to leave the city or factory."

It is clear that there are at least two countertendencies in the experiences of the people in the area: One, a decline in persisting farms, and the other, a new movement of people onto rural places and old farms for nonfarm purposes. Under these conditions, it is unavoidable that since the factual materials in the report are not segregated into these two diverse processes, they present a confused picture and fail to serve as testing evidence of the existence or the nature of the basic processes of human experience.

A similar confusion of divergent processes under a too rigid maintenance of the problem formulation is evidenced in an Illinois study published in 1939.[29] In this report the problem is formulated in terms of "the efforts of a relatively dense population to secure a livelihood by continuously cultivating soils best adapted for pasture, forest, and recreational uses." This interpretation is adhered to throughout the analysis of data, even up to the point where it is reported as "a startling condition" that "is difficult to believe" that a high percentage of rural relief applicants had no livestock of any kind and that over half of them had less than one tillable acre apiece.

These and other items which can be found in the report suggest that an analysis of the actual experiences of the people might have brought out the existence of various groups of families with characteristic backgrounds, experience, and problems requiring a refinement in the delineation and solution of the problem situation. Although the New Hampshire study failed to rearrange its analysis so as to expose and explore these sequential elements, it does take them somewhat into account in the conclusions; but in this Illinois report, these different patterns of experience are not only confounded in the evidence but are not taken into account in the conclusions.

[29] V. B. Fielder and D. L. Lindstrom, *Land Use and Family Welfare in Pope County* (Illinois Agr. Expt. Sta. RSM-8, 1939).

In the southern Appalachian region, Kentucky[30] and Tennessee[31] studies were published in 1937 and 1939. Each of these included a land classification technique as well as a sample procedure for farm survey. In both these studies the problem is formulated in terms of the decline of employment in mining and lumbering coupled with the increase in population on land not suitable for farming. Thus, in both studies, the nonagricultural character of the source of the problem is recognized; yet in both cases land classification techniques are based on a distinction between farm or forestry use. Consequently, a large portion of the occupied places are put in nonoccupancy categories, although the Kentucky study at least recognizes the contradiction between the classification definitions and the nature of the existing occupancy.

In both studies the farm schedule data are grouped not only by land class categories but also, though separately, by other categories such as lay of the land, tenure, or size of farm. Since these groupings are treated separately and for the entire sample, it becomes impossible to segregate groups of people by their patterns of experience so that the application of constructive solutions can be visualized. The Kentucky report concludes that the main emphasis must be on encouraging the people to make the best of their resources where they are, and the Tennesse study leaves to a local planning committee the task of figuring out how 18,000 people can be supported in a county that has room for only 10,000.

Three bulletins published in Indiana and Missouri between 1940 and 1942 illustrate the extent to which research in land utilization in the late 1930's tended to apply given techniques and problem formulations to problem situations in the highlands without any first-hand evidence to clarify the actual process of experience of the people in such areas. In all these reports the situation is described as one of declining agricultural enterprise and misuse of the land, and all of them have their objectives stated in vague terms: "to describe and analyze land use and related problems . . . to present factual information which may aid in the formulation of public policies," "to formulate general conclusions regarding adapted types of land uses," "to focus attention" on "social and economic maladjustments to improper land use."

A report on Reynolds County in the Missouri Ozarks is almost

[30] W. D. Nicholls, J. H. Bondurant, and Z. L. Galloway, *Family Incomes and Land Utilization in Knox County* (Kentucky Agr. Expt. Sta. Bull. 375, 1937).

[31] J. E. Mason and E. L. K. Gruehn, *An Analysis of Pertinent Social and Economic Factors Affecting Land Use in Overton County, Tennessee* (U.S.D.A. Land Economics Report 4, 1939).

exclusively a report on local government organization and operation with a few facts and figures on soil types, forestry, grazing, and crop production.[32] A bulletin on Martin County in the southern Indiana highlands presents some historical census data, an agricultural-nonagricultural classification map of the county, and a long discussion of local government problems.[33] A study of Callaway County, Missouri, is based on soil and building condition surveys from which the county is divided into eight areas which are described in general terms.[34] Nowhere in these studies is there any direct evidence of those experiences of rural families that are stated or implied in the problems or in the conclusions of these inquiries.

A study in Ross County, Ohio, issued in 1940 and based on data obtained from field interviews, is noteworthy because of the problem statement and the selection of sample areas.[35] Hilly parts of the county are recognized as failing to "supply a satisfactory living for the inhabitants," yet the standard suggestion to retire such areas to forest is regarded as "an incomplete answer." Therefore, the "study is intended to add a few more details of information . . . to show how the people in the poor land areas can be aided." No specific hypothesis, however, is given.

In the study, five areas were selected specifically to represent "distinguishing characteristics which justify some separate analyses," for there are "relevant circumstances which cause each area to be considered different from the others." But in the analysis, which is mainly an item-by-item summary comparison of the five areas, the significance of the areal differentiation fades, while the existence of different family experiences within the areas becomes noticeable. Thus, for example, four of the five areas are similar in the degree to which the families are dependent on the land for their living, although this factor is important in the analysis. On the other hand, there are families in each area that get less than 10 per cent or more than 75 per cent of their income from the land.

The same observation applies consistently through the various

[32] R. J. Silkett, *Land and Fiscal Problems in Reynolds County, Missouri* (Missouri Agr. Expt. Sta. Res. Bull. 324, 1940). The increased attention by land economists to local government problems was noted in Chapter V in respect to research in the cutovers. A further indication of this trend is a later Missouri report: F. A. Clarenbach, *Needed Local Government Reorganization in Ozark Land Use Adjustment Areas* (Missouri Agr. Expt. Sta. Res. Bull. 331, 1941).

[33] J. B. Kohlmeyer, *Major Land Use Problems in Martin County, Indiana, with Suggestions for Programs and Policies* (Purdue Agr. Expt. Sta. Bull. 453, 1940).

[34] A. M. Meyers, Jr., and C. H. Hammar, *Land Use Experience in Callaway County, Missouri* (Missouri Agr. Expt. Sta. Res. Bull. 346, 1942).

[35] H. R. Moore, *Hill Land and People in Ross County, Ohio* (Ohio Agr. Expt. Sta., Dept. of Rural Economics, Mimeo. Bull. 125, 1940).

items in the analysis. In respect to the conclusions, had the analysis grouped families with similar patterns of experience and ambitions, the recommendations of the study—for example, the suggestion that a large acreage of woodland under sustained-yield forestry be developed on each land holding—would be more convincing than they are.

Reminiscent of Minnesota Bulletin 180 (1918), is the fact that in this 1940 Ohio report data of each separate interview schedule are given in the report with the comment that "It is highly desirable to get a perspective of the combination of circumstances associated with each land holding and the family occupying it." But once more the task of ordering these combinations of circumstances into significant experimental categories is left to the reader instead of being a part of the procedure of the inquiry.

Attention may appropriately be given to a 1941 report on a back highland town in New Hampshire.[36] This study was conducted by sociologists and is referred to as a community culture study; yet its roots are planted in the land utilization changes in the area. The report has no formal presentation of data in the way of statistical tables, but among five formal graphic presentations there is a land utilization map, a graph of the distribution of landholdings, and a map of residential shifts.

Because of the orientation toward land use of the history of this town, the study clearly must be regarded as a part of the literature of social science research on problems of land utilization. What difference or contribution, then, results from this community culture study by sociologists? The answer hinges both on the research procedure used and on the viewpoint of the researchers who made the study.

On the positive side, there is no doubt that this inquiry makes a contribution that is not found in land utilization studies of similar areas in that (1) it shows a better understanding of the area as a whole—as a social organism—than is usually found, and (2) it demonstrates a better understanding of the historical processes—of what is going on. All through the report some distinction is kept among the various families—the farmers, the nonfarm laborers, the summer people, the urban dwellers whose patterns of experience with reference to the town are essentially different. Also, there is a constant attempt to give explanations for the developments described. As a result, the reader develops a sense of understanding the time sequence

[36] Kenneth MacLeish and Kimball Young, *Culture of a Contemporary Rural Community, Landaff, New Hampshire* (U.S.D.A. Rural Life Studies, 3, 1942).

and the interrelations among the process patterns which constitute the community's life.

On the other hand, the report cannot be regarded as a problem-solving type of inquiry. Perhaps because the report is a cultural study, it is pointed toward social association and characteristics of personal attitude, yet there is no problem definition and no hypothesis. An early section entitled "Major Problem of Adjustment and Maladjustment Now Facing the Community" refers only to the emigration tion of the town's young people and concludes that this is a "reasonable adjustment" to the "unalterable situation" that the town "has no future." Similarly, the final section of the report looks favorably upon "the disintegration of the hill towns" which is described in these terms: "The old farm houses will be abandoned or restored by summer visitors, and the old English names will be replaced by new ones . . . but the people . . . will only have gone to other places." Clearly, this study has something to contribute to the advancement of research in land utilization in the techniques of observation and recording social information, but it does not help in the definition of problems or the formulation of significant conclusions.

Brief reference should be made to the last two pre-World War II studies of land utilization in the eastern highlands. Appearing in 1942 was a report on Knott County, Kentucky, dealing with land use changes over the period 1929–1939.[37] The striking thing about this bulletin is the fact that it fails to seize upon an opportunity really to test a proposal made years before in another Kentucky mountain study which has already been referred to as exceptional.

Inasmuch as the Laurel County, Kentucky, report of 1932 urged the development of roads as a strategic point in solving the land utilization problems of the area, and inasmuch as Knott County was a mountain district without a single hard-surfaced road across it until 1933, the Knott County study might well have been centered on testing experimentally the conclusion of the Laurel County report, especially in view of the fact that data had been collected in Knott County in 1929. Although the 1942 Knott County bulletin makes two very brief qualitative references to the apparent effect of the new county road on outside incomes and school attendance, the bulk of the report is an item-by-item comparison of summaries of 1929 and 1939 farm records of average size of farms, tenure, land cover, crop production, livestock, and the like. These data are in line with the non-

[37] R. H. Allen and C. G. Deaton, *Trends in Land Use and Related Changes in Knott County, Kentucky, 1929–1939* (Kentucky Agr. Expt. Sta. Bull. 428, 1942).

problematic statement of the purpose of the study to "add to existing knowledge . . . rather than to develop plans or recommendations."

A 1943 study was also issued for two West Virginia counties that had been surveyed over a decade before in an earlier land utilization study.[38] The first part of this report is like the Ross County, Ohio, study in that the data are separated for three selected sample areas, but no particularly significant results attach to this geographical differentiation. The later sections, which analyze the forestry problem, contain some interesting illustrations of experimental research procedures.

Notable is the attempt "to test the feasibility of forest-farm homesteads under private property" in view of the fact that such homesteads had become a common recommendation for the solution of land utilization problems. In this study, a small area was found with a number of existing elements that came close to the concept of a forest homestead community. Working from this actual base, the study constructs an analysis of the hypothesis that if this community were altered so as to consist of forest homestead units, the area would be established as a stable community. Combining actual facts of experience with synthetic projections through an analysis of the process of converting the community, the study convincingly establishes the conclusion that the suggested proposal would not satisfactorily relieve the situation.

This type of analysis is clearly superior as a scientific procedure to the customary forestry analyses, in which a few simple figures on forest growth serve as basis for broad optimistic statements of expected forestry benefits, without taking into account the condition of the people located in the forestry environment.

Urbanization of the Countryside

It has been pointed out in the foregoing reviews of land utilization studies of the cutover and eastern highland areas that the assumption of a two-pronged choice between commercial agriculture and forestry has frequently persisted, even while evidential materials have suggested the actual or potential importance of other alternatives in the utilization of the area. In fact, one of the most striking characteristics of land utilization research in the cutover and in the highland areas is that, in the main, concepts and problem formulations that were crystallized before 1930 were carried over and adhered to all through

[38] E. C. Weitzell and L. F. Miller, *Forest-Land Utilization in Nicholas and Webster Countries, West Virginia* (West Virginia Agr. Expt. Sta. Bull. 309, 1943). Cf. footnote 24 above.

the middle 1930's without regard to the tremendous impact of the industrial unemployment which became widespread after 1929.

Interestingly enough, however, agricultural economists have given some attention to the play of nonfarm influences on the economy of nonurban areas. This attention came to a peak in 1935 and 1936.

URBAN INFLUENCES ON FARMS

Even before World War I, rural social scientists gave some attention to the impact of urban centers on the surrounding countryside. As early as 1913 and 1915, farm management surveys were made near Louisville, Kentucky, to show the differences in land values, farm product prices, farm expenses, types of farming, and farm incomes on commercial farms located at varying distances from the city.[39]

A similar inquiry appears in a bulletin issued in Indiana in 1932. It compares the changes between 1920 and 1930 on 60 farms near the Chicago region with the changes on 65 farms that were some distance removed but were nearly similar in soil and climatic characteristic.[40]

These studies, however, are clearly farm management studies for they deal with persisting units continuously devoted to farming and are concerned only with the internal management practices on those units.

URBAN RELATIONS OF FARM PEOPLE

Another line of inquiry that began before World War I was one which described the town and village trade and social relations of farm people. These investigations have almost uniformly been conducted by rural sociologists. At first these studies mainly concerned themselves with discovering the convergence of farm families' interests on various trade centers and the changes in such patterns of interest.[41] More recently, however, there has been more interest in depicting the differences not only in social and trade relationships but also in farming and other characteristics, at varying distances from a large urban center.[42]

[39] J. H. Arnold and F. Montgomery, *Influence of a City on Farming* (U.S.D.A. Bull. 678, 1918).

[40] Lynn Robertson, *Changes in Farming in Lake and Potter Counties, Indiana, as a Result of Nearness to Industrial Cities* (Purdue Agr. Expt. Sta. Bull. 365, 1932).

[41] As examples, see C. J. Galpin, *The Social Anatomy of An Agricultural Community* (Wisconsin Agr. Expt. Sta. Res. Bull. 34, 1915); C. C. Zimmermann, *Farm Trade Centers in Minnesota, 1905–1929* (Minnesota Agr. Expt. Sta. Bull. 269, 1930); J. F. Page, *Relations of Town and Country Interests in Garfield County, Oklahoma* (Oklahoma Agr. Expt. Sta. Bull. 194, 1930); P. L. Landis, *South Dakota Town-Country Trade Relations, 1901–1931* (South Dakota Agr. Expt. Sta. Bull. 274, 1932); A. M. Paxson, *Relations of Open Country Families of Onandaga County, New York to Socio-Economic Areas, Villages, and Cities* (Cornell Agr. Expt. Sta. Bull. 584, 1934).

[42] See J. F. Thaden, *The Lansing Region and Its Tributary Town-Country Communities* (Michigan Agr. Expt. Sta. Special Bull. 302, 1940).

RURAL RESIDENCES OF NONFARMERS

Before World War I there was some interest in questions concerning the location in the country of the residences of persons employed in nonfarm occupations. In 1918 the farm management office of the U.S. Department of Agriculture reported the results of an investigation of the value of rural homesteads to operatives in southern cotton mills; and in 1924 a study of the living costs of rural nonfarm workers was reported in New York State.[43]

In the spring of 1929 a land economist made a canvass of all rural homes in one Massachusetts town to determine the extent to which the families combined off-farm work with rural residence and farming activities. This geographically complete but otherwise brief investigation led to a further exploration of the nature of part-time farming in two Massachusetts urban areas.[44] The publication of this report, which estimated that half the state's farms were operated on a part-time basis, coincided with the growth of industrial unemployment throughout the nation. Among the many current suggestions for alleviating unemployment was the idea of encouraging the unemployed to live where they could cut living costs and provide some of their own food supply. Soon studies "to determine the possibilities" of providing a family "an adequate standard of living" and a "supplement to their income" by "small-scale farming" were started in Maryland, and surveys of the status of part-time farmers were set in motion near San Francisco and Los Angeles, California, and Rochester, Elmira, and Albany, New York, and the northwestern urban centers of Indiana.[45]

It is characteristic of most of these early studies that the data used as evidence are data which describe the condition of the part-time farm families who were interviewed in field surveys. The California report is designed specifically to show the income possibilities of part-time farming as reflected in the condition of existing part-time farmers. The same is true of the Maryland report except that it shows

[43] W. C. Funk, *Value of a Small Plot of Ground to a Laboring Man* (U.S.D.A. Bull. 602, 1918); C. V. Noble, *The Cost of Living in A Small Factory Town* (Cornell Agr. Expt. Sta. Bull. 431, 1924).

[44] David Rozman, *Part-Time Farming in Massachusetts* (Massachusetts Agr. Expt. Sta. Bull. 266, 1930).

[45] W. P. Walker and S. H. DeVault, *Part-Time and Small Scale Farming in Maryland* (Maryland Agr. Expt. Sta. Bull. 357, 1933); R. L. Adams and J. L. Wann, *Part-Time Farming for Income* (California Agr. Expt. Sta. Bull. 581, 1937); L. B. Tate, *The Rural Homes of City Workers and the Urban-Rural Migration* (Cornell Agr. Expt. Sta. Bull. 595, 1934); Kenneth Hood, *An Economic Study of Part-Time Farming in the Elmira and Albany Areas of New York, 1932 and 1933* (Cornell Agr. Expt. Sta. Bull. 647, 1936); Lynn Robertson, *The Economic Significance of the Non-Farming Rural Population in Northwestern Indiana* (Purdue Agr. Expt. Sta. Bull. 388, 1934).

average data for full-time farms by comparisons. The Elmira and Albany, New York, report also has this individual point of view except that the data include comparisons with urban living costs.

The Indiana survey is the only early study that attempts to assess the impact of the growth of rural nonfarm residences on the areas in which it takes place. Like the first phase of the Massachusetts investigation, the Indiana study covers a fully blocked rural area. It asks "What is [the] economic significance" of the growth of nonfarming rural homes "in the areas to which the movement takes place?" This report is based on very scanty information, yet it is interesting to note that attention centers on the degree of competition created for regular farmers, the local market created by the presence of nonfarm families, their effect on farm labor costs, public relief, and school costs, and their tax contribution. While some of the other studies mention these issues in their conclusions, in the Indiana study, these points are the center of the analysis. But it should also be noted that, whereas early in the report the nonfarming families are divided into four apparently quite different groups in respect to the experience of the families (city commuters, local rural workers, depression migrants, and retired persons), these groups are abandoned in the rest of the analysis. In view of the nature of the social issue to which the study was presumably addressed, the retreat from the first subsort of the cases is tantamount to discarding an excellent tool for dissection of the existent problem.

After 1932, as was pointed out in Chapter II, a federal program was in existence to help relocate stranded or unemployed populations. As a part of this program, a survey of existing part-time farms was inaugurated in 33 states. These data were later published as bulletins in eight states.[46] In nearly all instances the reports follow a fairly consistent pattern of describing statistical averages and frequencies of the items obtained on these standardized schedules. For the most part, the data were obtained only from part-time farms located close to urban centers. Uniformly, the purposes of the resulting publications

[46] F. L. Morison and J. H. Sitterley, *Rural Homes for Non-Agricultural Workers — A Survey of Their Agricultural Activities* (Ohio Agr. Expt. Sta. Bull. 547, 1935); I. G. Davis and L. A. Salter, Jr., *Part-Time Farming in Connecticut: A Preliminary Survey* (Storrs Agr. Expt. Sta. Bull. 201, 1935); G. W. Kuhlman, T. J. Fippin, and E. J. Niederfrank, *Part-Time Farming in Oregon* (Oregon Agr. Expt. Sta. Bull. 340, 1935); B. H. Pubols, *Part-Time Farming in Washington* (Washington Agr. Expt. Sta. Bull. 316, 1935); R. E. Wakeley, *Part-Time and Garden Farming in Iowa* (Iowa Agr. Expt. Sta. Bull. 340, 1935); Merton Oyler and W. W. Rose, *Part-Time Farming in Four Representative Areas of Kentucky* (Kentucky Agr. Expt. Sta. Bull. 358, 1935); Merton Oyler, W. W. Rose, and W. D. Nicholls, *Part-Time Farming by Negroes Near Lexington, Kentucky* (Kentucky Agr. Expt. Sta. Bull. 365, 1936); M. M. Daugherty, *Part-Time Farming in New Castle County, Delaware* (Delaware Agr. Expt. Sta. Bull. 199, 1936); F. V. Smith and O. G. Lloyd, *Part-Time Farming in Indiana* (Purdue Agr. Expt. Sta. Bull. 410, 1936).

are given as "discovering the extent and characteristics of part-time farming" and "presenting facts which may be helpful." One report specifically states that it "does not attempt to analyse underlying conditions nor to resolve the many problems of the field."

In addition to the bulletin presenting the data collected on a nationwide scale under the Division of Subsistence Homesteads program, a few state-initiated studies have been reported in more recent years. Like the earlier studies, most of these reports consist of statistical summaries of the characteristics of the part-time farm families which were interviewed for information.[47] A careful analysis of 24 reports of part-time farming studies published between 1930 and 1938 reached this conclusion: "As a rule, data have been secured only from schedules taken on part-time farms. There is a strong tendency for the researchers then to present the items from the schedule in terms of averages subsorted by some uniform, but not necessarily consequential, factor such as net farm income, total acres per farm, or minor civil division. . . . Statistics . . . should be developed as the answer to a question is followed, rather than as a set tabulating outline dictates. The groupings of data should be made by dominant motivations, patterns of behavior, sets of economic relationships, or outstanding problems, rather than by subsorts of isolated statistical items."[48]

RURBANIZATION

There is no good reason for classifying the foregoing studies as land utilization investigations inasmuch as they are concerned almost exclusively with the description of an existing form of social units as they are found. Some of the studies were conducted by farm manage-

[47] H. E. Moore and O. G. Lloyd, *The Back-To-The Land Movement in Southern Indiana* (Purdue Agr. Expt. Sta. Bull. 409, 1936); M. E. John, *Part-Time Farming in Six Industrial Areas in Pennsylvania* (Pennsylvania Agr. Expt. Sta. Bull. 361, 1938); C. F. Reuss, *Social Characteristics of Part-Time Farmers in Washington* (Washington Agr. Expt. Sta. Bull. 380, 1939); C. A. Boonstra and H. Jackson, *Part-Time Farming in A Rural-Industrial Area of Louisiana* (Louisiana Agr. Expt. Sta. Bull. 333, 1941); H. L. Hawley, *Small Agricultural Holdings in Two Industrial Areas in Indiana* (Purdue Agr. Expt. Sta. Bull. 460, 1941); C. F. Reuss, *Back to the Country — The Rurban Trend in Washington's Population* (Washington Agr. Expt. Sta. Bull. 420, 1942); W. R. Gordon, *Satellite Acres* (Rhode Island Agr. Expt. Sta. Bull. 282, 1942); R. C. Headington and J. I. Falconer, *Part-Time Farming in Three Land Use Areas of Columbiana County, Ohio* (Ohio Agr. Expt. Sta., Dept. of Rural Economics, Mimeo. Bull. 152, 1942); A. A. Asadorian, *Recent Resettlement in Rural Rhode Island* (Rhode Island Agr. Expt. Sta. Bull. 293, 1944). Also see R. H. Allen *et al.*, *Part-Time Farming in the Southeast* (U.S. WPA Res. Mono. IX, 1937), and review of the same by L. A. Salter, Jr., in *Journal of Farm Economics*, vol. XX, 3 (Aug. 1938): "It would seem that a tabulation rather than the problem at hand had been analyzed," insofar as standardized tables constitute the evidence.

[48] L. A. Salter, Jr., and L. F. Diehl, "Part-Time Farming Research," *Journal of Farm Economics*, vol. XXII, 3 (Aug. 1940).

ment experts and some by rural sociologists. These part-time farming studies were of particular interest to land economists mainly because part-time farms were associated with a new trend and form of land occupancy the characteristics of which were not known, as were those of other common forms of use and occupancy units found in rural areas.

The interest of land economists in respect to these movements would be expected to center in the process by which land devoted to a certain pattern of use was transformed to a different pattern of use and in the problems involved in making the adjustment. Attention would then be given to the landed economy as a whole and to the process of change in that situation. Actually, a few such studies have been issued. A start had been made on such studies in the area of Philadelphia, Pennsylvania, and Portland, Oregon, when World War II required their abandonment.[49]

In 1935 a report was made of a study that covered a whole valley in Connecticut and included an analysis not only of the farms, part-time farms, factories, and schools but also of urban people in the valley cities in respect to their interests in becoming part-time farmers.[50] This study was aimed at probing the hypothesis that if the establishment of part-time farms were facilitated by public action, then the "social conditions in the area" would be improved. Somewhat allied to the objectives of this Connecticut investigation is a much later report of a "critical appraisal" of nine of the actual experiments of the Division of Subsistence Homesteads in trying to facilitate part-time farming. This study, undertaken chiefly by sociologists, was altered and shortened by the outbreak of war; but it stands as an indication of how actual experiences in society can be regarded by social scientists as experiments for scientific inquiry.[51] The idea of the inquiry was to work out a careful statement of the ends-in-view in the subsistence homesteads undertaking and of the relation between the actions taken and the attainment of those ends, through case studies of the projects.

Undoubtedly the most extensive studies of the processes of change as nonagricultural activities move into a farming area are those made by Connecticut sociologists.[52] Although these studies tend to put more

[49] See L. F. Diehl, "Problems of Suburbia," *Land Policy Review*, vol. IV, 8 (Aug. 1941); Frederick Arpke, "Land Use Control in the Urban Fringe of Portland, Oregon," *Journal of Land and Public Utility Economics*, vol. XVIII, 4 (Nov. 1942).

[50] L. A. Salter, Jr., and H. D. Darling, *Part-Time Farming in Connecticut — A Socio-Economic Study of the Lower Naugatuck Valley* (Storrs Agr. Expt. Sta. Bull. 204, 1935).

[51] Russell Lord and P. H. Johnstone, eds., *A Place on Earth* (U.S.D.A., 1942).

[52] N. L. Whetten and E. C. Devereux, Jr., *Studies of Suburbanization in Connecticut, 1. Windsor* (Storrs Agr. Expt. Sta. Bull. 212, 1936); N. L. Whetten and R. F. Field,

stress on such factors as membership in group activities than might be the case were the projects conducted by land economists, nonetheless they clearly deal with a land utilization process. In each case, however, the studies are descriptive of what is taking place, and are intended "to indicate some of the social and economic adjustments which are resulting" in the hope of contributing "to a more adequate understanding of the suburban movement."

INDUSTRIALIZATION

A somewhat different type of land utilization process is that engendered by the establishment of an industrial use of land in an area previously without industry. As early as 1928, agricultural economists issued a report in which a study of rural industrialization was made by comparing two areas: one in which a manufacturing city of 12,000 population had developed in the preceding decade and one in which no industrial development had ever taken place.[53] It should be noted that this report does not contain an analysis of the processes of change in the industrialized community, but only a comparison of the before and after effects of the process. Thus, it poses no solutions as to what to do to get what results, but is admittedly a "first step" to "awaken interest, provoke discussion, and aid in developing ideas." The accomplishment of this task is seen not just in the orderly arrangement of a variety of data from a number of different sources but also in the final section, which lists eight specific topics for further analysis.

A more direct analysis of th process of industrial and related land utilization developments in a rural area is a recent study of Henry County, Virginia.[54] This bulletin is worth notice because it shows that the greater part of the work was originally concerned with the application of the New York technique of land classification survey to the county. In the final report, however, the emphasis is turned to the processes of change in land utilization. Data relevant to this latter point are relatively scanty, but it is clear that in it are the most fruitful parts of the work.

It is emphasized that the land classification which was made "makes no distinction as to whether the land designated as not

Studies of Suburbanization in Connecticut, 2. Norwich (Storrs Agr. Expt. Sta. Bull. 226, 1938); N. L. Whetten, *Studies of Suburbanization in Connecticut, 3. Wilton* (Storrs Agr. Expt. Sta. Bull. 230, 1939). See also W. R. Gordon and G. S. Meldrum, *Land, People, and Farming in a Rurban Zone* (Rhode Island Agr. Expt. Sta. Bull. 285, 1942).

[53] C. E. Allred and J. C. Fitch, *Effects of Industrial Development on Rural Life in Sullivan County, Tennessee,* University of Tennessee Record, vol. V, 3 (1928). Also see T. B. Manny, *Rural Factory Industries* (U.S.D.A. Circular 312, 1934).

[54] W. L. Gibson, Jr., and S. Bell, Jr., *Land Utilization in Henry County* (Virginia Agr. Expt. Sta. Tech. Bull. 93, 1944).

adapted to agriculture lies along the agricultural-forestry fringe or the rural-urban fringe. Furthermore, it makes no distinction as to whether the land adapted to agriculture can be put to a more intensive use through ru-urbanization." These comments certainly would seem to admit the relative usefulness of such classification work in an area characterized as one where "Industrialization has created ru-urbanization and part-time farming" so that "the site location of the land with respect to the industries and the highway system . . . becomes a strategic factor in determining its utilization . . . and adjustments in land use . . . lie not only along an agricultural-forestry fringe, but also along a rural-urban fringe." The argument that appears best to justify this work is that it was done "as a part of a larger program of inventorying the State's rural land resources according to their adaptability for agriculture."

On the positive side, the significant fact is that after making some farming summaries by land classes, the report turns to a wholly different basis of analysis in which there is a separate discussion of the importance, function, and trends in respect to each of nine types of land utilization units found in the county: full-time commercial farms, full-time subsistence farms (with and without nonfarm work), rural residences, part-time farms (residential, occupied abandoned farms, subsistence part-time farms, and commercial part-time farms), and nonfarm woodlands.

The great difference between this report and most others which apply the same land classification technique is that in this case the technique was not allowed to become an a priori solution of the problem. Consequently, new problem concepts, which actually outmode the concept for which the classification technique was devised, were developed. Although the classification work still accounted for the bulk of the investigational work so that the reformulated problems were not thoroughly analyzed, the conclusion of this report at least points to the next need as that of conducting "an exhaustive research program of the land use in the rural-urban fringe area of the county."

RECREATIONAL DEVELOPMENT

Another form of urban thrusts into rural areas that has attracted recent attention is the development of recreational land uses in non-urban areas. Mention has already been made of the failure of early land utilization studies in the Lake States cutovers and in the old northeastern highlands to include the development of private recreation as a new form of land use and occupancy. The study of recreational developments by rural social scientists began in 1932, since which time six reports have been issued. Like the early part-time farming studies, these are mainly single-use investigations—that is, they are concerned with existing recreational units rather than with the process of transformation from other uses to recreational use. Also, as might be expected, the reports are only exploratory in that they consist of "an inventory of real estate . . . that is used for recreational purposes," a classification of recreational unit types, and an estimate of the volume of business created by tourists and visitors.[55]

The recreational studies should be singled out for brief comment because they deal more completely with the process of transformation of land utilization in a rural area as recreational uses of the land develop. A Missouri study treats an area in which a large artificial lake was created for power purposes.[56] Although the study is only an "exploratory investigation of the nature of recreational developments," it brings together a wide variety of observations which focus upon a final series of three recommendations for more intensive analysis and one for positive action.

The other, an inquiry conducted in three New Hampshire towns, is the most thorough recreational process study yet reported.[57] In it is included information concerning the previous uses, the "motives and process of purchase" of land for recreational use, the relations between recreational and other types of use, and the future plans and intentions of the users of these lands. Also included is information concerning the nature of other land occupiers and other land uses, as well as an analysis of local government problems insofar as they are affected by the recreational development.

This study succeeds in achieving a dynamic element in its analysis, not by the actual handling of the quantitative data, but by the qualitative discussions which it contains and by the organization of the report in a kind of temporal order of topics. Also, the comparisons of

[55] G. S. Wehrwein and K. H. Parsons, *Recreation as A Land Use* (Wisconsin Agr. Expt. Sta. Bull. 422, 1932); David Rozman, *Recreational and Forestry Uses of Land in Massachusetts* (Massachusetts Agr. Expt. Sta. Bull. 294, 1933); W. O. Hedrick, *Recreational Use of Northern Michigan Cut-Over Lands* (Michigan Agr. Expt. Sta. Special Bull. 247, 1934); N. L. Whetten and V. A. Rapport, *The Recreational Uses of Land in Connecticut* (Storrs Agr. Expt. Sta. Bull. 194, 1934); W. R. Gordon and B. E. Gilbert, *Recreation and the Use of Land in Washington County* (Rhode Island Agr. Expt. Sta. Bull. 258, 1937).

[56] J. R. Snipe and C. H. Hammar, *Economic Aspects of Recreational Land Use in the Lake of the Ozarks Area* (Missouri Agr. Expt. Sta. Bull. 448, 1942). Also see a study of the uses made of recreational facilities developed on a federal submarginal land purchase area in Maine: E. J. Hiederfrank and C. R. Draper, *Use of Recreation Sites Developed on Federal Sub-marginal Land Purchase Areas in Maine* (Maine Agr. Expt. Sta. Bull. 280, 1940).

[57] J. C. Blum, *Land Utilization in New Hampshire, Part II. Summer Homes and The Rural Economy* (New Hampshire Agr. Expt. Sta. Bull. 344, 1942).

data among the three sample towns are subordinated to comparisons between two major categories of recreational land unit types — waterfront and open country properties.

As no hypothesis is formulated for testing, the New Hampshire report remains in the category of an exploratory inquiry. But the purpose is given as an attempt "to analyze the forces lying behind the summer home development and to indicate the nature of the social relationships and problems resulting therefrom." And the sorting of data is dominated by such concepts as that "the segments of the population which perpetuate this demand, and the channels through which personal desires are satisfied, is essential to an understanding of the effects . . . on the . . . economy of rural areas." Consequently, the study comes to "conclusions and recommendations" which cover eight pages of print and which are pregnant with suggestions as to existing and potential stages in the recreational development process at which changes can be instituted "to produce the maximum benefits to the community."

The Subhumid Regions

In terms of areal extent and intensity of interest, the problems of land utilization adjustment in the western part of the United States are among the most pressing land economics problems of the nation. Yet, in comparison with the problems previously discussed, these western problems have received relatively little formal study by social scientists. There are several probable reasons for this state of affairs.

In the first place, the West is the youngest part of the nation, and except for the Pacific Coast states has a sparse population and relatively little accumulated wealth. It therefore does not have the resources and the institutions to apply social research to the extent that is true in many eastern states. In the second place, most land problems of the West are directly administrative agency problems. A high proportion of the western lands are usable only for grazing purposes and most of these lands are actually owned and administered by federal departments. Similarly, the larger proportion of federal forest, park, and Indian lands are in this area. To a large degree, therefore, much of what would be done in a private land economy by public research agencies is done in the West as a regular part of the administrative work of public land administration offices. Even the development of private crop farms in the West is usually dependent upon an irrigation system, which in turn may be a project of the federal government. As a result, these agencies obtain information and make analyses which would be regarded as highly useful basic data for social science research, but the data are intended for internal administration or for legislative presentations and reports and cannot be safely regarded as a substitute for independent exploratory research.

In part because of this administrative dominance in western land problems and partly for other reasons, western land and water problems have been chiefly handled by biological scientists and engineers. Forestry and grazing problems have been treated mainly by foresters, agronomists, and animal husbandrymen, and irrigation problems by engineers and agronomists. In these areas, physical and technical knowledge have been necessary for proper administration, and since the economy has been largely dependent on administrative action, there has been but scant opportunity for the development of a specially trained corps of social science personnel. Finally, arising out of the complex just described, the land utilization problems of the West are highly charged with political implications, precisely because of the large hand that government has had in such resource development as the West has experienced. Under such circumstances, even though independent social science research may be most urgently needed, there is likely to be difficulty in establishing it.

In the present study, intensive analysis will not be applied to western land studies because of the indicated scarcity of formal reports on land economics research. Furthermore, the writer is not on familiar ground in respect to western problems; and although the literature of these regions has been carefully reviewed, no fundamentally different pattern of research has been found other than that already discussed in respect to the land utilization problems of other regions. At other points in this review, however, reference is made to western research dealing with land classification, tenure, and taxation.

GRAZING-FARMING ADJUSTMENTS

It is well known that the history of the Great Plains has been pocked by instances in which crop farmers have gone out on the semiarid and arid lands to establish homesteads in periods of good rainfall, only to be forced to evacuate the area in the next period of drought. Despite the drama of these experiences, they have attracted scant study of a careful nature.

Early attention, and always predominant research attention, to agricultural settlement of arid lands that would otherwise be used for grazing occurs in farm management studies.[58] Even the first work of the Division of Land Economics, in what was supposed to have been its part in a "comprehensive study of the agriculture of the Great Plains region . . . begun in 1920," was not issued until 1927; and it is but a summary of the business records of 196 dry-land farms for one year, plus some data on the financial progress made by those farmers since the time of their settlement.[59] In this report the only interesting methodological note is an early page warning of the dangers in interpreting averages per farm when enterprise summaries are made among farms some of which do not have the enterprise in question.

Over 15 years later, the Division of Land Economics issued another dry-land farm study which also is a summary of the business records of existing units.[60] Here again, there is no indication of any special land economics contribution, in the sense of studying the processes of adjustment between major land uses. For example, the report concludes with a recommendation for the expansion of grazing lands for most farmers and of crop land for some farmers. But this is the point at which land utilization inquiry should commence. It is just this impact on the conditions of landholding and use outside the units needing the adjustment that would be critical for a land utilization analysis.

Also, the occasion for this study was the fact that a group of crop farmers had stayed on the land despite widespread expectation that they would be forced to give up. In pursuit of this inquiry, a statistical comparison is given of "the financial condition of farmers upon entering the area and their condition in 1940." But following the observation that some had gained and some had lost, the report flatly states that "What happened between the two dates . . . to cause the varying fortunes . . . is a subject beyond the scope of this analysis."

All the land economics work on the grazing-farming frontier has not, however, been limited to orthodox farm management analyses of existing units. In the middle 1930's techniques of land use planning were employed in studies of farming-grazing adjustment areas in the

[58] E.g., F. S. Harris and A. D. Ellison, *Dry-Farming in Utah* (Utah Agr. Expt. Sta. Circular 21, 1916); O. R. Mathews, *Dry Farming in Western South Dakota* (U.S.D.A. Farmers' Bull. 1163, 1920); M. L. Wilson, *Dry Farming in the North Central Montana Triangle* (Montana Agr. Ext. Ser. Bull. 66, 1923); L. H. Hauter, A. L. Walker, and O. V. Wells, *A Five-Year Economic Study of 125 Farms in Curry and Roosevelt Counties, New Mexico* (New Mexico Agr. Expt. Sta. Bull. 186, 1930).

[59] E. O. Wooton, *Settlers' Progress in Dry-Land Farming in Eastern New Mexico* (U.S.D.A. Circular 4, 1927).

[60] O. E. Goodsell, *An Economic Appraisal of Dry-Land Farming on the Zuni Plateau, New Mexico* (U.S.D.A., 1943).

northern Great Plains.[61] These studies, based primarily on maps of operating unit patterns, land ownership units, land classifications, together with summaries of farm and ranch schedules, indicate the complexity of the materials that must be dealt with in such areas. The studies are aimed primarily at the delineation of areas that ought to be shifted from crop farming to grazing, and their final conclusions therefore are only that stated percentages of reduction in crop acreages are necessary and that fewer families than are in the areas can be supported by the resources of those areas.

In later land economic studies attention turned to the possibilities of public action. One study in Montana analyzes the experience of a grazing district organization as an experiment in adjustment in farming-grazing land utilization, a North Dakota study deals with a resettlement experimental project,[62] and a South Dakota study deals with the possible adaptation of rural zoning to the problems of the semiarid plains.[63]

Mention must also be made of two social studies of Kanas communities. One of these, a case study of Haskell County, Kansas, represents a careful review of the in- and out-migrations in semiarid areas referred to previously.[64] It is stated that "The fundamental purpose of this study" is to determine "what happens to the social institutions and human relationships in a community that is compelled to make drastic alterations . . . because of drought and depression." Elsewhere it is said that the study substantiates the "hypothesis . . . that social changes associated with drought tend to follow a definite sequence pattern." In this study, then, excellent as it is as an exposition of the general processes of land utilization adjustment in the area, no proposal for action was to be checked and tested. The report therefore is revealing in respect to the nature of the problems but inconclusive in regard to possible solutions. The same might be said of a

[61] John Muehlbeier and R. E. Johnston, *Some Land Use Problems in Northwestern South Dakota* (U.S.D.A., 1937); Stanley Wilner, *Land Use Problems in Southwestern North Dakota* (U.S.D.A., 1937). For a statewide summary of dry-land farming adjustments, see E. A. Starch, *Readjusting Montana's Agriculture VII, Montana's Dry-Land Agriculture* (Montana Agr. Expt. Sta. Bull. 318, 1936).

[62] J. H. Marshall and S. W. Voelker, *Land Use Adjustment in the Buffalo Creek Grazing District, Yellowstone County, Montana* (U.S.D.A. Report LE-6, 1940); J. P. Johansen, *One Hundred New Homesteads in the Red River Valley, North Dakota* (North Dakota Agr. Expt. Sta. Bull. 304, 1941).

[63] R. J. Penn and C. W. Loomer, *County Land Management in North-western South Dakota* (South Dakota Agr. Expt. Sta. Bull. 326, 1938); R. J. Penn, W. F. Musbach, and W. C. Clark, *Possibilities of Rural Zoning in South Dakota* (South Dakota Agr. Expt. Sta. Bull. 345, 1940); and M. H. Taylor and R. J. Penn, *Management of Public Land in North Dakota* (North Dakota Agr. Expt. Sta. Bull. 312, 1942).

[64] A. D. Edwards, *Influence of Drought and Depression on a Rural Community: A Case Study in Haskell County, Kansas* (U.S.D.A. Social Research Report VII, 1939).

community culture study of Sublette, Kansas, as has been said of the Haskell County report—and the community culture study of the old highland town in New Hampshire.[65]

GRAZING CONFLICTS

The problems of land utilization adjustments between grazing and crop farming are difficult to differentiate from adjustments between different types of grazing or ranching units. In some of the reports referred to in the preceding section, it is not always entirely clear whether the past existence of crop farm units has merely confounded the problem of securing adequate ranch units or whether crop farm units have persisted with consequent hardships to the farmers as well as difficulties for the ranchers. Failure to make this distinction eminently clear is not merely a matter that causes confusion to the uninformed readers of these reports. It is also an illustration of the common but fundamental error which this study has repeatedly pointed out: that social science research above all must reveal and maintain distinctions between basically different patterns of purposes, actions, and consequences as they are experienced by the people under observation, if the results of the research are to have standing as experimental evidence and usefulness in the solution of social problems.

A second complicating factor in a treatment of land utilization conflicts among competing forms of grazing land uses is that they hinge so largely on control over rights in land as to be almost indistinguishable from land tenure studies.

One of the earliest ranching management studies was published by the U.S. Department of Agriculture in 1915.[66] Early in this report it is stated that although the report is "mainly concerned with the details of the purely physical basis of the industry, the factors of control and relations to other industries are so closely connected with any proper kind of management that they must be considered somewhat at length." When the same investigator published his next bulletin in 1922, he was a member of the Division of Land Economics and the title of his report was *The Relation of Land Tenure to the Use of Arid Grazing Lands*.[67] In the same year a Texas study of ranch management referred to the field of ranch economics which embraced management, tenure, and utilization issues.[68]

[65] E. H. Bell, *Culture of a Contemporary Rural Community, Sublette, Kansas* (U.S.D.A. Rural Life Studies 2, 1942).

[66] E. O. Wooton, *Factors Affecting Range Management in New Mexico* (U.S.D.A. Bull. 211, 1915).

[67] E. O. Wooton, *The Relation of Land Tenure to the Use of Arid Grazing Lands of the Southwestern States* (U.S.D.A. Bull. 1001, 1922).

[68] Bonney Youngblood and A. B. Cox, *An Economic Study of a Typical Ranching Area on the Edwards Plateau of Texas* (Texas Agr. Expt. Sta. Bull. 297, 1922).

In subsequent years there have been a few studies of cattle or sheep ranch management issued in most of the western states. But nearly all of these reports emphasize the fact that range control problems, which lie outside the scope of an internal management study, are critical in the ranching economy.[69] Since 1931 there have been some publications centering on the grazing utilization and grazing tenure aspects of range problems.[70]

GRAZING-FORESTRY ADJUSTMENTS

One of the most controversial of the many conflicts over land utilization in the West concerns the competition between forestry and grazing in foothill areas of the Coastal and Sierra Nevada ranges. While grazing-forestry conflicts are found in greater or lesser degree in parts of the Ozark and southern Appalachian highlands and the Rocky Mountains, they have been even more intense in the western foothills.

It so happens that in many of the Pacific Coast areas where the grazing-forestry conflict is found, there are other major uses of the land which further complicate the situation. Among these are crop farming, which usually requires irrigation facilities, rural residences, part-time farms, and retirement homesteads, and other nonfarm uses and industrial enterprises. Here, then, are some extremely difficult problems, to which land economists have but recently turned their attention.

Three Oregon reports, two of which appeared in 1942, have been issued since 1940. Two of these studies are based primarily on a physical classification of the land with a view to indicating areas in which new settlement might be encouraged.[71] The third consists mainly of an analysis of the public measures available for regulating

[69] E.g., "The fact is well established by the prevailing system of operation that ownership of the range is the factor determining the system of range use employed. The problem of land ownership or control by lease is the influencing factor in stable operation." M. L. Wilson, R. H. Wilcox, and G. S. Klemmedson, *A Study of Ranch Organization and Methods of Range-Cattle Production in the Northern Great Plains Region* (U.S.D.A. Tech. Bull. 45, 1928).

[70] E. O. Wooton, *The Public Domain of Nevada and Factors Affecting Its Use* (U.S.D.A. Tech. Bull. 301, 1932); C. A. Brennen, *The Public Range and the Livestock Industry of Nevada* (Nevada Agr. Expt. Sta. Bull. 139, 1935); M. H. Sanderson and N. W. Monte, *Grazing Districts in Montana* (Montana Agr. Expt. Sta. Bull. 326, 1936); M. H. Saunderson, *Readjusting Montana's Agriculture, V, Economic Changes in Montana's Range Livestock Production* (Montana Agr. Expt. Sta. Bull. 311, 1936); Marion Clawson, Cruz Venstrom, and T. D. Phinney, *Range Lands of Northeastern Nevada* (U.S.D.A., 1938); G. H. Craig and C. W. Loomer, *Collective Tenure on Grazing Land in Montana* (Montana Agr. Expt. Sta. Bull. 406, 1943).

[71] R. E. Bell, J. Winter Smith, and Ray Deschamps, *A Reconnaissance Plan for Land and Water Use in Josephine County, Oregon* (U.S.D.A., 1940); and V. W. Baker and D. C. Mumford, *Land Settlement in the Willamette Valley* (Oregon Agr. Expt. Sta. Bull. 407, 1942).

settlement, grazing, and forestry.[72] In none of these studies is there a clear-cut analysis of the experiences of the owners or occupiers of the lands in question.

Three reports that deal with the California foothill complex have been published. The first, published in 1934, treats El Dorado County, the second deals with one part of Yuba County, and the third covers a large northern Sierra Nevada foothill area that encompasses the areas described in the two earlier reports.[73]

In terms of the complexity of land uses, it would be hard to find a more difficult situation than that dealt with in the El Dorado and northern Sierra Nevada area reports. In view of this fact the weakness of the conclusions is understandable, despite the indications that a great deal of work by several cooperating specialists and agencies was carried on in this general problem area. Yet it should also be noted that predominantly the new facts given in the report are in the form of elaborate maps. The text is primarily composed of general discussions of each of the major types of land utilization with some figures from secondary sources and only occasional indications of new materials. This report indicates that mapping techniques, however accurately applied and on however grand a scale, go but a short distance toward helping to solve complex land utilization problems. By comparison the relatively unpretentious report on Yuba County makes a great deal more headway by selecting the one and most important issue of grazing versus forestry, and by attempting to analyze the specific hypothesis that if rural zoning were used, then the forestry-grazing conflicts could be resolved and subsidiary benefits also enjoyed.

IRRIGATION DEVELOPMENTS

Land economists have not had much of a hand in research dealing with irrigation, despite the fact that almost any irrigation enterprise requires a considerable measure of social action in order to be established. Soon after the federal government became directly involved in irrigation development under the Newlands Act of 1912, Richard T. Ely was invited to Washington to work on the economics of irrigation. His visit did not last long, however, and a manuscript which he

[72] M. L. Upchurch, *Problems and Institutional Factors Affecting Land Use in a Northwest County* (U.S.D.A., 1942).

[73] David Weeks, A. E. Wieslander, and C. L. Hill, *The Utilization of El Dorado County Land* (California Agr. Expt. Sta. Bull. 572, 1934); H. E. Conklin, David Weeks, and R. B. Wertheimer, *The Possibilities of Rural Zoning in the Sierra Nevada Foothills, A Study in the Livestock-Forest Area of Yuba County, California* (U.S.D.A., 1942); David Weeks, A. E. Wieslander, H. R. Josephson, and C. L. Hill, *Land Utilization in the Northern Sierra Nevada* (California Agr. Expt. Sta. Special Pub., 1943).

had prepared for federal publication was never issued but lies in the files of the Wisconsin Historical Society as page proof of an unpublished report.

During the first twenty years of federal irrigation enterprise, the agricultural economists of the U.S. Department of Agriculture made farm management studies which dealt with alternative practices on irrigated farms as existing enterprises.[74] These studies, taking the existing units as given, could deal with the internal organization of irrigation farms without raising any general questions as to the process of experience that irrigation farmers went through or the feasibility of putting arid lands under the ditch.

After the organization of the Division of Land Economics, work was undertaken on the hitherto neglected land utilization aspects of irrigation. These efforts did not result in a series of studies of local areas, but they did result in the private publication of the only book on the subject of irrigation economics.[75]

Even to date there are comparatively few studies, other than farm management surveys, of irrigation farms. In 1937 Nebraska issued a bulletin on the economic benefits of a contemplated irrigation project in which it was indicated that "it is the purpose of this study to outline the agricultural and economic benefits that will accrue to the Platte river drainage area because of irrigation." [76] Whether this statement poses a question for scientific inquiry needs no answer.

Two recent reports are based largely on a land classification technique.[77] Both are intended to present data on the relative quality of the land so that interested persons will have more data on which to base their own decisions.

The most elaborate study by rural social scientists deals with the Yakima Valley in Washington. For this valley, a series of experiment station bulletins has been issued.[78] The series includes a good deal of

[74] E.g., E. H. Thompson and H. M. Dixon, *Profits in Farming on Irrigated Areas in Utah Lake Valley* (U.S.D.A. Bull. 117, 1914).

[75] R. P. Teele, *The Economics of Land Reclamation in the United States* (Chicago, 1927).

[76] Frank Miller and H. C. Filley, *Economic Benefits of Irrigation from the Kingsley (Keystone) Reservoir* (Nebraska Agr. Expt. Sta. Bull. 311, 1937).

[77] G. T. Blanch and C. E. Stewart, *Utilization of Irrigable Land in the Reservation Area of Uinta Basin, Utah* (Utah Agr. Expt. Sta. Bull. 303, 1943); B. H. Pubols, *Prospective Farming on the Columbia Basin Irrigation Project* (Washington Agr. Expt. Sta. Bull. 456, 1945).

[78] E. B. Hurd and H. F. Hollands, *Economic Conditions and Problems of Agriculture in the Yakima Valley, Washington: The Agriculture and Its Setting* (Washington Agr. Expt. Sta. Bull. 377, 1939); E. B. Hurd and H. F. Hollands, *The Yakima-Tieton Irrigation District* (Bull. 393, 1940); Wallace McMartin, *Part V. The Sunnyside Division of the Yakima Project* (Bull. 428, 1943); A. E. Orr, *Part VI. The Irrigation Project of the Yakima Indian Reservation* (Bull. 430, 1943). Another Yakima study, but in a different

information relating to the general history of the valley and the irrigation districts in it, but the central purpose of the bulletins is "to assemble pertinent information" that either may serve as a base for "any program of improvement" or that will "be useful to the farmer" in improving his economic condition.

Although these reports put a good deal of emphasis on farm management materials, they also include histories of the various projects in the valley as well as analyses of land tenure, changing patterns of utilization and holding, and land value trends. Also, each of the studies includes fairly complete maps of ownership and operating units, land cover, farm types, physical land classifications, and the like. In all the studies, main emphasis in the conclusions turns on reorganization of the pattern of land use and tenure, district consolidation, reassessment of lands, or other group actions.

The Yakima-Tieton report may be referred to as an illustration of a preliminary analysis. The only purpose given is that of assembling information that may be useful in program formulation. The core of the report contains several land maps of various types, a little quantitative tabular data, and material describing the evolution of the present land pattern. The conclusions of this inquiry include some well-stated hypotheses. For instance, "Operating costs of some units could be reduced if scattered holdings could be exchanged for contiguous lands." Or, "if repayment contracts . . . should require high payments . . . in years of high income, there would be a tendency to reduce indebtedness and keep land costs down." "If the cost of . . . an irrigation system is assessed according to the paying ability of the land, repayments . . . are likely to be maximized." The materials in the report suggest these hypotheses but are not adequate to test them as conclusions. It should be noted, however, that well-formulated hypotheses are proper conclusions for studies of the preliminary type.

One of the most thorough examinations of irrigation settlement is a study of new settlers on an Oregon project.[79] This report covers a great deal of information and results in many suggestions. Its purpose is to find, "as a result of experience" on one project, what can be suggested to advance the success of settlement on other projects. This

series, deals with the adjustment of laborers who migrated to the valley. See C. F. Reuss and L. H. Fisher, *The Adjustment of New Settlers in the Yakima Valley* (Washington Agr. Expt. Sta. Bull. 397, 1941), also issued as *New Settlers in Yakima Valley, Washington* (U.S.D.A. Migration and Settlement on the Pacific Coast Report 8, 1941).

[79] C. P. Heisig and M. Clawson, *New Farms on New Land* (U.S.D.A. Migration and Settlement on the Pacific Coast Report 4, 1941).

comparatively careful study may be referred to for an illustrative point. In the conclusions, three pages are devoted to suggestions for relating credit extension to housing needs and farm development needs. But in the evidence of the report, although there are discussions of credit experience, housing conditions, farm development, and incomes, each of these is treated separately. The conclusion referred to is surely suggested by the data; but it could have been tested had the report subclassified the cases in subseries patterns of type of housing, size of debt, and income experience. As it is, one knows there are high and low income settlers, well- and poorly-housed settlers, and settlers with much and little debt; but there is no evidence that reveals the condition of each settler with respect to all of these points within his own experience.

The foregoing review does not include reference to a number of specialized types of irrigation development studies. Particular mention should be made of a series of studies of water-supply organizations made late in the 1920's when irrigation districts and water companies were in difficult financial straits.[80] Also, there have been a few general historical reports based on secondary data, reports which deal with historical development in the usual sense of that term.[81] There have also been some legislative summaries of irrigation policies[82] and a few technical economic studies of irrigation economics.[83]

Other Land Utilization Problems

The number of possible types of land utilization problems other than those already referred to is extremely large. So long as there are developing needs or attempts to establish a type of major land use on areas occupied by units whose land requirements differ from the new use, problems for land utilization research will arise.[84] So far

[80] E.g., W. A. Hutchins, *Irrigation District Operation and Finance* (U.S.D.A. Bull. 1177, 1923); *Financial Settlements of Defaulting Irrigation Enterprises* (U.S.D.A. Circular 72, 1929); *Mutual Irrigation Companies* (U.S.D.A. Tech. Bull. 82, 1929); *Commercial Irrigation Companies* (U.S.D.A. Tech. Bull. 177, 1930); *Irrigation Districts, Their Organization, Operation and Financing* (U.S.D.A. Tech. Bull. 254, 1931); G. E. P. Smith, *The Financial Rehabilitation of Irrigation and Drainage Districts* (Arizona Agr. Expt. Sta. Bull. 144, 1933).

[81] E.g., R. J. Smith, *The California State Land Settlements at Durham and Delhi*, University of California Hilgardia, vol. 15, 5 (Oct. 1943).

[82] R. P. Teele, *Land Reclamation Policies in the U.S.* (U.S.D.A. Bull. 1257, 1924); R. B. Wertheimer, *Legislative and Administrative History of Acreage Limitations and Control of Speculation on Federal Reclamation Projects* (Bureau of Agricultural Economics, 1943).

[83] David Weeks and C. H. West, *The Problem of Securing Closer Relationship Between Agricultural Development and Irrigation Construction* (California Agr. Expt. Sta. Bull. 435, 1927).

[84] See J. D. Black, ed., *Research in Agricultural Land Utilization* (Social Science Research Council Bull. 2, 1931), p. 4, for a distinction between major and minor uses.

attention in this country has been mainly attracted to those previously discussed, but there are many others.

Because of the amount of time and space that would be necessary to describe the background and then to discuss methodological issues for the various minor types of land utilization problems on which some research has been done, the following reference to each type of situation will be very brief.

FLOOD CONTROL

In connection with flood control projects there are various aspects of major land utilization adjustments that have been made. A Missouri and a Mississippi study have dealt with the problems involved in the purchase of farm land to be flooded for flood control purposes and the relocation of the families so dislocated.[85] And a Louisiana report deals with the complex problems facing a community in an area which has been left open to backwaters.[86]

DRAINAGE

The land utilization adjustments involved when unused land is drained in order to put it into agricultural use have received even less attention in land utilization research than has irrigation. The Division of Land Economics has conducted one study of drainage district administration[87] and some of the cutover lands of the South referred to in Chapter V are open to settlement because the land has been drained. For the most part, however, drainage studies, like irrigation studies, have been primarily dealt with by engineers.

WILD LIFE

The utilization of land for wild life purposes is very often in serious conflict with the devotion of land to other major uses. This area of conflict, though important, has received scant attention from land economists. Because wild life uses are commonly public uses whereas other use of the same land is within a private economic unit, the opportunity for conflict and the need for resolution of such conflicts is very great. The beginning of a research attack on this type of prob-

[85] E. A. Wilkening and C. L. Gregory, *Planning for Family Relocation* (Missouri Agr. Expt. Sta. Bull. 427, 1941); R. R. Nichols and M. B. King, Jr., *Social Effects of Government Land Purchase* (Mississippi Agr. Expt. Sta. Bull. 390, 1943). Also see R. B. Wertheimer, *Flood-Plain Zoning* (California State Planning Board, 1941).

[86] T. L. Smith and S. E. Grigsby, *The Situation and Prospects of the Population in the Black River Settlement, Louisiana* (Louisiana Agr. Expt. Sta. Bull. 319, 1940).

[87] R. D. Marsden and R. P. Teele, *Economic Status of Drainage Districts in the South in 1926* (U.S.D.A. Tech. Bull. 194, 1930).

lem has just been made in a recent bulletin which discusses some of the issues in a general way.[88]

SUBSURFACE EXPLOITATION

Ordinarily there is no major conflict between the development of surface and subsurface resources, but recently the technological developments in strip mining of coal have created problems in certain areas to which a little research attention has been given.[89] Also, there has been growing awareness of the problems that oil and gas leasing and drilling can cause for surface land users. These conflicting uses will undoubtedly grow in terms of the research attention they will draw in the future, but to date they represent a very minor part of the field.

SPECULATIVE TOWNSITES

One of the most fascinating problem areas in land utilization reported on is the Pine Barrens of New Jersey. The basic process involved is one which has occurred in various areas at certain times (for example, Florida, 1925). Speculative promoters sell rural lands in urban lot sizes on the appeal that a new town is to be founded. The difference in New Jersey is that the bases of the appeal vary from project to project; and because of the vastness of the Pine Barrens and their close proximity to great centers of population, the problem is perennial rather than sporadic.

This problem has been the subject of inquiry of two land economics studies.[90] New Jersey Bulletin 665 ought to be singled out for comment because it illustrates an original approach. The study represents a historical case study in which the evidence runs back to 1888 and up to the present in such a way that it becomes impossible to say where the past history stops and the current analysis begins. Furthermore, all through the report, the attempt to follow through the actions of the people requires the collection and arranging of a wide variety of types of evidence all of which is focused on the outcome of the project which is used as a case.

[88] J. P. Miller and R. B. Powell, *Game and Wild-Fur Production and Utilization on Agricultural Land* (U.S.D.A. Circular 636, 1942).

[89] H. R. Moore and R. C. Headington, *Agriculture and Land Use as Affected by Strip Mining of Coal in Eastern Ohio* (Ohio Agr. Expt. Sta. Rural Economics Mimeo. Bull. 135, 1940). See also James Salisbury, Jr., and L. A. Salter, Jr., "Subsurface Resources and Surface Land Economics," *Journal of Land and Public Utility Economics,* vol. XVII, 3, 4 (Aug. and Nov. 1941).

[90] A. T. M. Lee, *Land Utilization in New Jersey: A Land Development Scheme in the New Jersey Pine Area* (New Jersey Agr. Expt. Sta. Bull. 665, 1939); J. F. Hauck and A. T. M. Lee, *Land Subdivision in the New Jersey Pines* (New Jersey Agr. Expt. Sta. Bull. 701, 1942).

FARMING PATTERNS

It has already been noted that in arid and semiarid areas, because of the magnitude of the land factor, management adjustments often depend upon the making of major land utilization adjustments. Under certain circumstances in the humid regions, especially under the impact of momentous technological changes, there may be new readjustments within a given system or type of farming that are on such a scale as to involve far-reaching rearrangements in the whole pattern of land utilization and tenure. When such changes take place incidentally and over a long period of time, they are not likely to attract attention as a social question, but when many such changes take place in a short period of time, the resulting problems may create noticeable community readjustments as well as personal difficulties. Recently, there have been some indications that under modern technological developments this stage may be reached in some parts of the Middle West. There have been, consequently, a few studies which have begun to probe this phenomenon in terms of its origins and its resultant effects on the local agricultural economy.[91]

Recapitulation

HISTORICAL

Land utilization research in areas other than the cutover regions came relatively late in the development of land economics. During the period of settlement and development prior to World War I, there was very little economic research work on the problems of getting land into use by dry farming and irrigation.

With the relative decline of agriculture after World War I, attention in the East was attracted to farm land abandonment in the northern highland areas and to the lack of adjustments in the southern Appalachians. In New York State early work on abandoned areas preceded the adoption of a large-scale state reforestation program and an accompanying program to classify land that should be considered for reforestation. From this latter program many states adopted the map-making technique and issued land classification

[91] J. R. Hays, *Relation of Character of Farming Units to Land Management in Two Townships in Indiana* (Purdue Agr. Expt. Sta. Bull. 450, 1940); B. R. Hurt, E. C. Young, and L. Robertson, *Land Use Adjustments Needed on Farms in Deer Creek Township, Cass County, Indiana* (Purdue Agr. Expt. Sta. Bull. 466, 1942); W. H. Pine, *Area Analysis and Agricultural Adjustments in Nemaha County, Kansas* (Kansas Agr. Expt. Sta. Bull. 305, 1942); R. C. Headington and J. I. Falconer, *Changes in the Size of Farming Units in Three Land Use Areas of Hancock County, Ohio, 1937–1940* (Ohio Agr. Expt. Sta. Rural Economics Mimeo. Bull. 148, 1942); M. J. Peterson and G. H. Aull, *Land Utilization and Agricultural Adjustment in Edgefield County, South Carolina* (South Carolina Agr. Expt. Sta. Bull. 349, 1944).

reports on counties scattered throughout the Northeast and even in central and western United States.

Meanwhile, the Division of Land Economics promoted interest in the land abandonment phenomenon, and studies were undertaken in several northeastern states and in the highland areas of the central states. This work shows a joining of interests of the farm management approach and what may be termed a local geographical approach. The studies also reflect a growing interest in the public finance aspects of the land utilization changes that were the subject of concern.

In the depths of the depression of the early thirties there was a wave of research attention to the settlement in rural areas of urban people, particularly on land in the vicinity of large cities. Although there is evidence that this back-to-the-land movement also affected the more remote areas that had been in the process of abandonment (as was also noted to be true for the Lakes States cutovers in Chapter V), in general the research work in such areas shows little recognition of the back-to-the-land phenomenon. In other words, the depression settlement and the decadent area problems were examined in separate research compartments.

In respect to the problems of land use adjustment in the subhumid areas of the West, the paucity of land economics research is noticeable. It is particularly so in view of the fact that the settlement of arid lands has always involved serious questions of public policy. Although Ely and Taylor gave attention to the economics of irrigation development early in the century, and although Ray P. Teele worked on irrigation problems for years in the Reclamation Bureau and in the Division of Land Economics, the only substantial work is a book privately published by Teele just before his death.

The Division of Land Economics issued a few reports on the land utilization adjustment problems of grazing lands about 1920, but very little else was reported until after the Taylor Grazing Act of 1934 and the establishment of the programs of land use planning during the New Deal period. Since then there have been a few reports on grazing adjustment problems.

The relative inattention to land utilization problems of the subhumid regions is particularly noticeable in view of the fact that range management studies consistently stress the fact that the critical problems are those of general land use and tenure adjustment. Also, the large amount of range and irrigation investigational work done by physical scientists and engineers adds emphasis to the lack of work on these problems by social scientists.

With the extension of land economics research work on almost a state-by-state nationwide scale during the 1930's, there is noted a widening of the type of land utilization problems on which research has been done, including surveys of problems caused by speculative townsite promotion schemes and by exploitation of subsurface resources.

Finally, it is found (as was the case for the cutover regions) that toward the end of the period reviewed, rural sociologists began to turn their attention to the problematic situations on which land economists had been working. These workers have contributed by reemphasizing processes of human behavior and by undertaking work that recognizes action programs to be sources of experimental social data.

METHODOLOGICAL

The chief weakness of land economics research revealed in the work reviewed in this chapter is the freezing of problem formulation so that the objective becomes merely counting or classification work rather than the exploration of the sources of problems and possible modes of solution. Under this condition detailed and expensive work is sometimes done — without regard for its relevance to the situation to which it is applied. Also, pieces of information that might prove to be clues to possible suggestions for action are disregarded as exceptional rather than as tests of the initial concept of the problem or of the hypothesis.

On the other hand, another major source of difficulty is that surveys are conducted with no problem posed for inquiry, with the result that the data presented are not aligned as evidence of the consequences of alternative actions.

In many studies set arrangements of quantitative data are adhered to instead of being readjusted so as to promote the elaboration of the hypothesis and to reveal the underlying experience of the subjects of the study. In some instances these standardized groupings of data are maintained even when the materials as presented suggest that a different arrangement would be more appropriate to the problem at hand or would provide a more thorough test of the conclusions.

In several instances where there is a determined interest in exploring the problematic situation by breaking it down into patterns of experience and data are arranged to clarify these patterns, new suggestions of problems for further inquiry are posed or new ideas of remedial actions are suggested.

CHAPTER VII

Research in Landed Property

One of the two basic issues in the revolution in land policy at the turn of the century was the increasing difficulty of achieving ownership of farm land by farm operators. Interest in this land tenure problem figured in the early development of agricultural economics, particularly at Wisconsin. The research work done on land tenure problems represents the first real substance, in research, of what later came to be rural land economics.

Landlord and Tenant Relations

Ever since the census reports of 1880 and 1890 began to classify farms as tenant- or owner-operated, economists have produced a stream of journal articles summarizing and interpreting the trends in and geographical distribution of these most general farm tenure statistics.

WISCONSIN BULLETIN 198, 1910

In 1899 and 1901 Henry C. Taylor spent several months in England collecting materials for his thesis on the decline of farm land ownership. As part of his work he toured the country talking with tenants and landlords. After his return to the United States and the publication of his thesis, he obtained funds to travel in southern Wisconsin interviewing farm tenants and land owners about their renting problems. The results of this experience were published in 1910 in the first agricultural experiment station bulletin dealing with land tenure problems and one of the earliest of experiment station bulletins on the social side of agricultural problems.[1]

This Wisconsin bulletin is illustrative of the research approach previously noted in the northern Wisconsin settler study of 1920 by Ely and Hibbard, except that it is even more informal. In the introduction it is stated that there is need to find ways of resolving conflicts between landlords and tenants, so the "purpose of this bulletin is to summarize the experience of a large number of families who have talked freely with the writer on this subject."

[1] H. C. Taylor, *Methods of Renting Farm Lands in Wisconsin* (Wisconsin Agr. Expt. Sta. Bull. 198, 1910).

The report is devoid of any quantification of these experiences, but it is clear that in the survey, tenants and landlords were invited to specify the sources of conflict in their relationships and to state how the conflicts experienced by others were resolved in their experience. For instance, the report indicates that one source of difficulty arises in respect to colts raised by tenants. The bulletin states that this particular problem "has been solved in various ways," which are then explained.

This bulletin is important in respect to certain characteristics that relate to the methodological discussion in Chapter III. On one hand, it is as far removed from formal quantification as is possible. But on the other hand, the task in this study was seen as that of finding out what specifically were the felt problems which in sum created the general social question of dissatisfaction with landlord-tenant arrangements. This was accomplished by quizzing the people concerned. At the same time it was seen that in some other instances cases would be found in which these specific problems had been solved in experience. What this inquiry did was to treat the visited tenant farms as experiments for (1) exposing and refining the social problem that previously was only vaguely formulated, and (2) discovering ways in which these specific problems had been resolved in experience.

It should also be noted that the same person who was perplexed to begin with went to look into these experiments, probed them, and reported his observations. At this point, question may be raised as to whether this could have been done had there been a determination in advance of what the specific problems were and what standard factual data were needed from every farm that was visited. If, for example, the accounting of colts raised by tenants had not been known to be a problem by the investigator in advance of his field work, and if it were not a problem on the first farms visited but was a problem on, say, the eleventh farm visited, it would only be by chance that the materials taken on the first ten farms would constitute relevant evidence for this specific problem. But having run onto the issue on the eleventh farm, the investigator could in subsequent visits be sure to check whether or not this same form of the tenancy problem existed; and if so, why, and if not, why not. An important requisite for research — an open mind and a keen eye — is called for, not only in this, but in all types of research.

Without holding up this bulletin as a model of scientific perfection it is very important to note these attributes of it. When the report treats a certain specific problem and tells how it was overcome, the reader may not know on how many farms this occurred; but even though he would be better informed if he did know the number, he does know that in this group each and every farm had overcome the problem by the same process. All this contrasts with procedures in which the facts forming coherent wholes for this group of farms are blended in with facts for other farms so that the existence and the solution of the specific problem can at best only be hinted at by elusive differences in the group averages, and may escape entirely unobserved.

In 1911, the year following the Wisconsin report, the U.S. Department of Agriculture issued a little circular on farm tenancy which describes the tenancy system on one estate in Maryland with 56 tenant farms.[2] Here again, quantification is absent. While it is impossible to tell, for example, how many estates can be found like the one in question, on the other hand it is clear that where the stated conditions obtained, the stated outcomes were also experienced. In fact, the case was selected specifically because in this instance the phenomena usually construed as the undesirable results of tenancy were absent. So the investigators looked into this experiment to see why the usual results did not appear and whether there were corollary problems or concomitant results not apparent on the surface. Although this study was put into the farmers bulletin popular series, probably because either it was not regarded as scientific inquiry or it had directly perceivable significance for landlords and tenants, it should be noted that two years later, the department issued in its research series a farm management study consisting of a single case analysis.[3]

THE SURVEY METHOD INFLUENCE

Before the next study of farm tenancy was issued, George Warren's famous *An Agricultural Survey* had been published.[4] While Warren's report covered many more subjects than farm tenancy, it nevertheless had a section devoted to the topic. For this dissertation, however, the importance of Warren's bulletin is not what he published about farm tenancy, but the procedures he used.

The difference between Taylor's tenancy bulletin and Warren's survey bulletin is not that they did not both go out to interview farmers. Nor is it that they did not both regard farmers' experiences as

[2] J. W. Froley and C. B. Smith, *A System of Tenant Farming and Its Results* (U.S.D.A. Farmers' Bull. 437, 1911).

[3] M. C. Burritt and J. H. Barron, *An Example of Successful Farm Management in Southern New York* (U.S.D.A. Bull. 32, 1913).

[4] G. F. Warren and K. C. Livermore, *An Agricultural Survey* (Cornell Agr. Expt. Sta. Bull. 295, 1911).

ex post facto experiments—to use the modern sociological phrase of Chapin and Greenwood. In fact, the very first sentence in the Warren bulletin is, "Every farm is an experiment station and every farmer the director thereof."[5] But the introduction goes on to say that because any one farmer "is quite likely to attribute success or failure to the wrong cause," it is "by studying large numbers of farms [that] the real reasons and their relative importance usually stand out clearly." The statement goes on to discuss the accuracy of the survey procedure, not in terms of the dangers of wrong causes or the clarification of real reasons, but solely in terms of the mathematical accuracy of averages of quantified data. That this interpretation of the advantages or disadvantages of this research procedure came to be dominant is indicated in a later bulletin on the *Validity of the Survey Method of Research* by Spillman. Here the entire discussion of validity is in terms of the application of the law of averages to quantified data obtained from field interviews, as compared with that from cost accounts and plot experiments.[6]

What was not seen, but what is more important, is that Taylor searched for "experiment station" farms that were encountering like problems; and among these he mentally and then verbally put together those that used the same controls to get the same outcomes. What Warren did was to assume that standardized data from all experiment stations, even though they were conducting different experiments, would throw light on various issues so long as a sufficient number of farms were included.

The difference in research results is that whereas one does not know to what number of farms Taylor's problem solution may apply, one does not know to what problems many of Warren's numbers apply. Thus, the New York bulletin reports that cash-rented farms average 99 acres in size and pay an average of $186 in rent and "This amounts to $1.88 per acre or $2.66 per tillable acre," but "The rent per acre varied from $0.77 to $5.32." Still, in spite of this wide variance, there is not a word as to whether some rents are too high or too low or even whether the level of rents is considered to be any problem at all.

[5] Exactly the same introductory statement is developed in the same way in the agricultural survey of a different area made 21 years later by George Warren's son. S. W. Warren, *An Economic Study of Agriculture in Northern Livingston County, New York* (Cornell Agr. Expt. Sta. Bull. 539, 1932). It is also interesting to note that in 1932, E. H. Thompson comments on W. J. Spillman as one who "never lost sight of the fundamental fact that every farm was, in a sense, an experiment station." "The Origin and Development of the Office of Farm Management in the U.S. Department of Agriculture," *Journal of Farm Economics*, vol. XIV, 1 (Jan. 1932).

[6] W. J. Spillman, *Validity of the Survey Method of Research* (U.S.D.A. Bull. 529, 1917).

It is not implied that Warren's bulletin does not raise any other problems or that the analysis rests only on averages. In the analysis of farm profits, qualitative discussions and individual case summaries are given. But the important point here is that Warren's work set the direction of emphasis on quantitative completeness and accuracy and on mass averages and away from formulation of problems and experimental case experiences.

The quantified survey procedure has of course developed apace. Accumulated facts have piled up, usually in "deadly parallel columns."[7] Reactions against this development, how ver, have centered on its failure to use more refined statistical techniques[8] or to make greater use of deductive theoretical economic reasoning.[9] In the contrast between Wisconsin Bulletin 198 and Cornell Bulletin 295 one can see the difference even more clearly than in the contrast between Wisconsin Bulletin 318 and Minnesota Bulletin 180 (used in Chapter V), in emphasis on problems and on similarities in action and consequence patterns within observed cases.[10]

It is not altogether clear that Warren's techniques should have come to be regarded as the only survey method any more than Taylor's; the difference was in the quantification of the results of their farmer interviews rather than in any other respect, for they both obtained research data by interviewing farmers in the field.

As a matter of fact, through Galpin, Taylor introduced a third form of field survey procedure. When Galpin first sought in 1909–1910 to study the group contacts of rural families, he had the librarian in a New York town obtain data on the memberships held by owner and tenant families in that town. These data were summarized by Galpin by marking symbols on a map where the respective family lived.[11] This survey method was the first to use a geographic analysis

[7] G. W. Forster, "Discussion of Recent Developments in Research Method and Procedure in Agricultural Economics by H. R. Tolley," *Journal of Farm Economics*, vol. XII, 2 (May 1930).

[8] See Chap. III above, especially references to the Social Science Research Council handbook of 1928.

[9] See Forster, "Discussion," *Journal of Farm Economics*, vol. XII, 2 (May 1930); T. W. Schultz, "Scope and Method in Agricultural Economics Research," *Journal of Political Economy*, vol. 47, 5 (Oct. 1938), and "Theory of the Firm and Farm Management Research," *Journal of Farm Economics*, vol. XXI, 3 (Aug. 1939); also see Chap. III above. H. C. Taylor has written that since Warren's first survey, "Many men have since used the survey method of research . . . and many of them have put in a life-time with this method without getting a flash, but Warren had the genius to develop many principles of farm management out of the materials collected in the Tompkins County survey . . . I am convinced that no one else has ever got anything like [so] much out of the survey method as he did." "Early History of Agricultural Economics in the United States," Part I (Unpub. ms. in University of Wisconsin Agricultural Library, June 1941).

[10] Cf. Chap. III, "Research Procedures."

[11] C. J. Galpin, "The Story of My Drift Into Rural Sociology," Part II, *Rural So-*

of rural social science data. It led to the mapping of the trade and social contact areas of villages and cities and was the forerunner of map techniques later used in land utilization surveys.[12]

LEASE DESCRIPTIONS

With the rise in land values between 1910 and 1920, landlords of the increasing proportion of rented farms began to feel that adjustments should be made in leases since rental rates lagged behind the upward surge of land and other prices. Similarly, during the 1920's and 1930's, as farm prices and incomes dropped, there was a demand for further attention to leasing arrangements to protect tenants. Consequently, some attention has constantly been given to the various forms that rental contracts take, especially in respect to the division of farm receipts and expenses. This interest has clearly peaked, however, in periods of rapidly rising or rapidly falling farm incomes. To a very large extent this type of problem has been conducted by agricultural economists whose interests are in farm management rather than in other land economic questions. Furthermore, no other aspect of land tenure problems has received anything like the attention that has centered in bulletins on lease contracts.

By far the majority of the bulletins on leasing arrangements merely describe the contents of various types of rental contracts that are in existence.[13] Most of these reports are based on a mail survey of land-

ciology, vol. II, 3 (Sept. 1937). Also see C. J. Galpin, "The Social Agencies in a Rural Community," *First Wisconsin Country Life Conference* (Madison, 1911), pp. 12–18.

[12] See Chaps. V and VI above.

[13] Examples, omitting Extension Service publications, are: E. V. Wilcox, *Lease Contracts Used in Renting Farms on Shares* (U.S.D.A. Bull. 650, 1918); E. A. Boeger, *Rent Contracts in Typical Counties of the Wheat Belt* (U.S.D.A. Bull. 850, 1920); R. L. Adams, *California Farm Tenancy and Methods of Leasing* (California Agr. Expt. Sta. Circular 272, 1923); C. L. Holmes, *Drawing Up the Farm Lease* (Iowa Agr. Expt. Sta. Circular 87, 1923); H. E. Selby, *Farm Rental Terms* (Montana Agr. Expt. Sta. Circular 119, 1923); C. C. Taylor and J. J. Vernon, *Renting Farms in Virginia* (Virginia Agr. Expt. Sta. Bull. 249, 1926); F. I. Riddell, *Farm Lease Systems in Michigan* (Michigan Agr. Expt. Sta. Circular Bull. 102, 1927); W. E. Grimes, *The Stock Share Lease* (Kansas Agr. Expt. Sta. Circular 155, 1930); G. R. Johnson, *The Farm Tenant and His Renting Problem* (Missouri Agr. Expt. Sta. Bull. 315, 1932); Millard Peck, *A Plan for Adjusting Cash Rent to Changes in the Prices of Farm Products* (Iowa Agr. Expt. Sta. Bull. 295, 1932); C. E. Miller and W. O. Brown, *Farm Tenancy and Rental Contracts in North Dakota* (North Dakota Agr. Expt. Sta. Bull. 289, 1937); O. R. Johnson, *Acquiring Farm Ownership by Payments in Kind* (Missouri Agr. Expt. Sta. Bull. 378, 1937); W. D. Nicholls, *Share Leasing Contracts* (Kentucky Agr. Expt. Sta. Bull. 307, 1937); J. F. Timmons, *Landlord-Tenant Relationships in Renting Missouri Farms* (Missouri Agr. Expt. Sta. Bull. 409, 1939); H. C. M. Case and Joseph Ackerman, *Farm Leases for Illinois* (Illinois Agr. Expt. Sta. Circular 503, 1940); A. E. Orr, *Leasing Washington Farms* (Washington Agr. Expt. Sta. Bull. 385, 1940); R. L. Adams and W. H. Smith, Jr., *Farm Tenancy in California and Methods of Leasing* (California Agr. Expt. Sta. Bull. 655, 1941); E. D. Tetreau, *Arizona Farm Leases* (Arizona Agr. Expt. Sta. Bull. 179, 1942); E. B. Hill, *Father and Son Farm Partnerships* (Michigan Agr. Expt. Sta. Special Bull. 330, 1944).

lords or tenants or both to obtain descriptions of their existing leases. Some of the bulletins have been issued as extension circulars, others as research bulletins. Some are restricted to a single general type of lease or special plan. Sometimes they contain some elementary data showing the trend in the percentage of farm tenancy among farms and perhaps maps showing the distribution of various major types of leases. Usually, sample copies of lease forms are included.

None of these bulletins can be regarded as a research report; they are informational rather than investigational. Their objective, as stated in U.S.D.A. Bulletin 850, is to acquaint the tenants in one area with "the methods of renting in vogue" elsewhere. Where additional facts concerning the trend and distribution of tenancy are included, they are aimed, as North Dakota Bulletin 289 says, "to provide some material upon which intelligent discussion and constructive thinking about the problem of tenancy may be based."

INCOME FROM RENTED FARMS

A second group of bulletins is fundamentally styled like the first group, but in addition the reports include some analysis of the results of using different types of leasing systems. These bulletins, like the first group, formulate the problem within the landlord-tenant framework. As Minnesota Bulletin 178 states it, "It is the purpose . . . to consider the rental systems and leases now in use . . . No effort is made to go into the details of the evils of tenancy, to discuss its causes, or to suggest a remedy." The "urgent and immediate need is for better rental systems and this is the phase of the subject taken up in this investigation." [14]

In these bulletins, the landlord and tenant problem is further formulated toward two ends-in-view: (1) higher farm profits, and (2) allocation of income between landlord and tenant in proportion to costs borne by each.[15] The general procedure then is to compare average tenant incomes with average landlord returns with the schedules classified by lease type. In the more comprehensive studies, com-

[14] A. H. Benton, *Farm Tenancy and Leases* (Minnesota Agr. Expt. Sta. Bull. 178, 1918).

[15] O. G. Lloyd, *Farm Leases in Iowa* (Iowa Agr. Expt. Sta. Bull. 159, 1915); H. A. Turner, *Systems of Renting Truck Farms in Southwestern New Jersey* (U.S.D.A. Bull. 411, 1916); Benton, *Farm Tenancy;* W. E. Grimes, *Farm Leases in Kansas* (Kansas Agr. Expt. Sta. Bull. 221, 1919); B. H. Hibbard and J. D. Black, *Farm Leasing Systems in Wisconsin* (Wisconsin Agr. Expt. Sta. Res. Bull. 47, 1920); J. I. Falconer, *Methods of Renting Land in Ohio* (Ohio Agr. Expt. Sta. Bull. 348, 1921); A. H. Benton, *Cash and Share Renting of Farms* (North Dakota Agr. Expt. Sta. Bull. 171, 1924); B. H. Hibbard and Harold Howe, *The Farm Lease in Wisconsin* (Wisconsin Agr. Expt. Sta. Bull. 391, 1927); J. E. McCord, *Farm Tenancy and Lease Forms* (Pennsylvania Agr. Expt. Sta. Bull. 232, 1929); P. I. Wrigley, *Farm Tenancy in Pennsylvania* (Pennsylvania Agr. Expt. Sta. Bull. 383, 1933).

parisons may also be made of the proportionate expenses incurred by the landlord and tenant. There may also be some simple averages which compare yields, investments, or other items for the various rental types.

Referring to these reports, H. A. Turner says that "For the most part, current systems of leasing land have been challenged only in rather inconclusive ways, and the published information . . . is fragmentary."[16] The present review substantiates this statement. Not the least of the reasons why the average comparisons of farms rented under different types of leases are inconclusive is that the classification of large numbers of farms by lease type is often also a classification of these same farms on other important but hidden bases. That this common source of difficulty may have been present is inadvertently suggested in a study of tenancy types and types of farming in Iowa which shows that different leases are commonly used for different types of farming.[17]

A few bulletins that deal with landlord-tenant arrangements in slightly individualistic ways may be considered separately.

In Missouri Bulletin 167 (1920) special attention is given to the average rents paid per acre and the average landlord returns on investment for different crops on lands of different value levels.[18] This series of data repeatedly shows that average rents and average landlord returns are lower under cash than share agreements. But there are no data that provide an explanation for this phenomenon. For instance, it is said that "as farms become less fertile the owners . . . demand cash rent rather than share rent." Clearly a substantiation of this would require sequential data showing that on a number of farms cash leases had in fact replaced share leases as yields fell. The data presented raise many other similarly engaging questions; but since the tables represent broad factual generalizations of the relationships of facts drawn from various points in the processes of various experiences, they can only perform the function of suggesting these interpretations without grounding them as warranted assertions.[19]

[16] In J. D. Black, ed., *Research in Agricultural Land Tenure* (Social Science Research Council Bull. 20, 1933), p. 53. Also see Joseph Ackerman, "Status and Appraisal of Research in Farm Tenancy," *Journal of Farm Economics*, vol. XXIII, 1 (Feb. 1941).

[17] C. L. Holmes, *Relation of Types of Tenancy to Types of Farming in Iowa* (Iowa Agr. Expt. Sta. Bull. 214, 1923).

[18] O. R. Johnson and R. M. Green, *Renting Land in Missouri* (Missouri Agr. Expt. Sta. Bull. 271, 1930).

[19] It should be noted that the Missouri station for some 20 years after the publication of this bulletin continued in other reports to probe the same productivity-value phenomena. See B. R. Rawlings, Jr., and O. R. Johnson, *Relationship of Productivity of Farm Units and Their Ability to Pay Rent* (Missouri Agr. Expt. Sta. Res. Bull. 308, 1939). Also see Missouri Bulletins 315, 378, and 409 cited above.

For the most part, Virginia Bulletin 271 (1930) offers the same type of information as the previous bulletins reviewed: a summary of existing lease terms and a description of landlord and tenant returns.[20] In one particular, however, the study should be singled out. Reflecting the then new and popular use of statistical correlations, the study presents a graphic and statistical indication that "high returns for either the landlord or tenant were accompanied by high returns for the other." But instead of leaving this analysis of relationships at that point, there is a designation of 13 specific cases which fall outside the band on the chart within which most cases lie. These dots "represent farms on which the returns to one of the parties were more than twice as high as the returns to the other party relative to their averages." Picking these as problem cases the study continues: "An examination of the record secured from each of these farms reveals many of the reasons for the unequal division of return." A special summary of data, identified for each case individually, is given. The available information shows that some of these cases are easily explained by their leasing terms and others by abnormal management practices. For the rest, the report can only indicate apparent possible sources of difficulty.

This analysis illustrates generalization of relationships as serving the function of suggesting clues to the existence and analysis of problems that actually are found within the experience of the units of observation.[21]

Kentucky Bulletin 303 (1930) also centers on the amount and division of returns on rented farms.[22] The report contains data on this point in the area concerned for three different years. In addition, there are short paragraphs giving the average and percentage figures on tenant family expenses, moving expenses, living conditions, life insurance, literature read, memberships in organizations, and whether the attitude of the tenants was optimistic or pessimistic. Some school attendance figures are given to show that changes in school enrollment coincide with the tenant moving period. Presumably all these non-farm management data are given because they bear on the tenancy problem, but in the report they are not tied in with any problem

[20] R. A. Ballinger, *Stock Share Renting in Virginia* (Virginia Agr. Expt. Sta. Bull. 271, 1930).

[21] It should also be noted that in the study reported in Pennsylvania Bulletin 383, "Special inquiry was made to locate farms operated under unusual leasing arrangements." Such searches for the unusual rather than the representative are important if social scientists are to engage in research that is truly experimental. Clearly, if the customary actions give problematic results, it will be in unusual cases that suggestive ideas for solutions will be most likely found.

[22] W. D. Nicholls, *Farm Tenancy in Central Kentucky* (Kentucky Agr. Expt. Sta. Bull. 303, 1930).

statement, problem analysis, or any recommendation for action. They do meet the object of the study "to make available facts . . . bearing on the tenancy problem."

In addition to the fact that Maryland Bulletin 352 (1933) presents some types of factual data that are not found in most bulletins on farm leasing, it serves as a clear illustration of some points that are pertinent to this review.[23] The problem is formulated in terms of the shift of farming from crops to livestock and the "failure of the present leasing contract . . . to keep step" with changes in price levels, types of farming, and farming methods. In the body of the report it is maintained that the leasing agreement in effect in some areas "is not equitable," that some landlords are not "in a financial position to bring about . . . desirable changes," that present arrangements do "not give the landlords much inducement to make needed improvements." Nevertheless, the report concludes that "Present conditions do not seem to justify, or permit, any material changes in the leasing system as it exists in Maryland." Falling back on the U.S.D.A. Circular 437 of 1911, which gives the case report on one Maryland estate, this 1933 bulletin offers some "fundamental principles" that "should be kept in mind in making the rental agreement." But these principles have no direct relation to the ordering of data presented in the report of the study.

Since it is clear that a central issue in the Maryland problem is the provision of livestock facilities by the landlords, a classification of landlords into economic groups would seem to provide a significant lead. Actually, however, the groupings are described as an isolated procedure; no attempt is made to divide the farms surveyed into these groups nor are the groupings in any other way related to the analysis of tenanted farms. Rather, throughout the report, the data are grouped, averaged, and compared by counties.

Colorado Bulletin 451 (1938) introduces a new approach in that it compares landlord and tenant incomes over a period of time.[24] Unfortunately, the "same farmers do not continue throughout the 14 years" and "several variations in methods of leasing are included, hence the averages are not representative of anything but themselves." If this forthright statement demolishes any significant purpose that the tables in the bulletin are supposed to achieve, it still should be emphasized that at least there is data for one farm continuously, and

[23] W. P. Walker and S. H. DeVault, *Farm Tenancy and Leasing Systems in Maryland* (Maryland Agr. Expt. Sta. Bull. 352, 1933).

[24] R. T. Burdick, *Landlord and Tenant Income in Colorado* (Colorado Agr. Expt. Sta. Bull. 451, 1938).

apparently successfully, tenanted throughout the 14 year period. Fortunately, this farm is segregated so that the history of costs and returns is clear and the variations in crop yield, prices, and the like. In addition, the bulletin compares the 14-year average division of incomes and expenses on this one farm with a synthetic representation of what the division would have been under a lease form that had been recommended elsewhere. Although the quantitative data suggest that certain relations obtain between the lease provisions and the equity of division of farm returns, it is quite obvious in this study that the treatment of the one landlord-tenant case over a period of years gives the conclusions such substance as they have.

In 1942 appeared a Nebraska study of farm renters which goes beyond the previously limited area of tenant studies.[25] In the opening of this report the problem is formulated somewhat vaguely in terms of avoiding "unwise attempts to solve" the tenancy problem. Although this problem is not put in specific terms, the report includes an "evaluation of tenancy" and suggests "guiding principles in solving the tenancy problem."

The chief characteristic of the data obtained for this study as compared with landlord-tenant inquiries previously reviewed is that the tenants and landlords were asked not only for specific facts concerning their present status but also for their preferences in respect to the same matters. Taking these factors up one at a time, the report shows, for example, that nearly all share renters accept the terms of the rent as satisfactory, that, on the average, cash tenants prefer a rate lower than the existing one, and that the majority of tenants are satisfied with the livestock and grain storage facilities on their farms. The preference of tenants for more livestock is described as a "desire rather than a sound practical view" because of the limits of the feed supply.

In the "evaluation of tenancy," a shift in data is made to include farm records from farms in ten surrounding counties in addition to those of the survey county. Income data for five- and ten-year periods show that the average rate of landlord earnings is below the farm mortgage interest rate, and that tenants' incomes on small farms average less than that needed for family living. Following this, a synthetic analysis indicates that average tenant earnings were higher than they would have been had the tenants been owners with only a $5000 equity in their farms. Finally, some data on the changing of farms by tenants indicates that nearly half the moves are occasioned

[25] L. F. Garey, G. H. Lambrecht, and Frank Miller, *Farm Tenancy in Clay County, Nebraska* (Nebraska Agr. Expt. Sta. Bull. 337, 1942).

by the tenants' desire to move and that a number of tenants have long periods of residence on their places.

Although this analysis clearly brings a variety of new facts to bear on the tenancy problem, it is confined, as are most studies, to comparisons of lump sums. For example, granted that a majority of tenants are satisfied with their building facilities, if there is a problem to be solved with respect to buildings on rented farms, it will be necessary to segregate the tenants who do not have satisfactory facilities and to examine the landlord-tenant conditions on those farms to determine what, if anything, can be done to eliminate the problem.

Again, knowing that there are more tenant moves on resident landlord farms where tenants are not related to the landlord than on nonresident landlord farms, a solution of the problem will require, not only this quantitative comparison, but study of the resident landlords with steady tenants as well as a study of nonresidents with a high tenant turnover.

Lacking such analyses as are here suggested, the bulletin can conclude only with general recommendations that do not stem from the data of the report but from previous knowledge and belief.

AREA STUDIES OF TYPE OF TENANCY

In recent years, some descriptions of existing lease arrangements have given greater attention to the existence of practices that have been generally recommended as a part of desirable tenure conditions, and they have given greater geographic precision to descriptions of the prevalence of detailed provisions.[26]

Just before World War II, a project was started in a number of states to delimit and describe "type of tenancy areas." Apparently, the object of those who initiated this work was to reduce landlord-tenant problem analysis to an area basis, somewhat in the manner that studies of type of farming had attempted earlier for farm management problems.[27] The publications that resulted from these efforts offer a wide array of tenancy descriptions.[28]

[26] A. J. Englehorn, *Farm Tenure in Iowa, VI. Landlord-Tenant Relationships in Southern Iowa* (Iowa Agr. Expt. Sta. Bull. 372, 1938); D. M. Pettus, *Farm Rental Agreements in Caswell County, North Carolina* (U.S.D.A., 1940); C. E. Allred, B. H. Luebke, and L. D. Malphrus, *Landlord-Tenant Relations in Roane County, Tennessee* (Tennessee Agr. Expt. Sta. Rural Research Mono. 117, 1940); W. H. Scofield and H. C. M. Case, *Farm Leasing Practices in Illinois* (Illinois Agr. Expt. Sta., 1942); O. G. Lloyd, H. S. Morine, Jr., and J. R. Hays, *Principal Methods of Share Renting and Compensation for Unexhausted Improvements in Four Type-of-Farming Areas in Indiana* (Purdue Agr. Expt. Sta. Bull. 464, 1942).

[27] See Joseph Ackerman, "Status and Appraisal of Research in Farm Tenancy," *Journal of Farm Economics,* vol. XXIII, 1 (Feb. 1941).

[28] J. C. Elrod, *Types of Tenancy Areas in Georgia* (U.S.D.A., 1941). Also see J. C. Elrod, *Graphic Summary of Farm Tenancy in Georgia* (Georgia Agr. Expt. Sta. Bull.

In general, the procedure in the type of tenancy studies is to assemble on maps data describing the distribution of selected characteristics and then to delineate areas within which these characteristics occur in greater or lesser density. The chief factors used are: percentage of tenancy, kinship between tenant and landlord, kinds and amounts of rents paid, value of farms. In some cases, information was sought on the degree of management supervision provided by the landlord. In the Louisiana work the information is reduced to this single factor and the type of tenancy areas are distinguished by the prevalence of independent, semi-independent, and supervised tenants. In all reports, after the areas are marked on a map, a verbal description of general conditions is given for each area.

Just what these type of tenancy area studies have contributed is difficult to say. Their stated purposes are to overcome "lack of knowledge of tenancy conditions by geographic areas" or "to present the geography of farm tenancy." With such a lack of attention to problem formulation as exists in these studies, it is only possible to note that they do not constitute full scientific inquiries. It is even doubtful whether they represent valuable introductions to scientific inquiry since none of them conclude with formulations of problems for research. In the Ohio report the summary states that "variation in tenure characteristics and methods of renting between individual farms are apt to be more pronounced than are the distinguishing characteristics of the areas as a whole. However, since one of the purposes of the study was to make out areas in the State having somewhat similar tenancy characteristics, five such tenancy areas have been delineated." While such a conclusion exhibits excellent devotion to duty, it hardly amounts to much of a contribution from compiling facts on type of tenancy. The Indiana report does go so far as to suggest that standard lease forms should be drawn up for each area; but the areas referred to and used in the Indiana analysis are type of farming areas rather than type of tenancy areas.

LEGAL ASPECTS OF LANDLORD AND TENANT RELATIONS

Another series of bulletins has been issued from various states

210, 1941); I. W. Moomaw, *Farm Tenancy Areas in Ohio* (Ohio Agr. Expt. Sta. Rural Economics Mimeo. Bull. 144, 1941); F. J. Ramsey and Harold Hoffsommer, *Procedure in Delineating Types of Tenancy Areas in Louisiana* (U.S.D.A., 1941), and *Farm Tenancy in Louisiana* (U.S.D.A., 1941); J. R. Motheral, *Types of Farm Tenancy Areas in Texas* (U.S.D.A., 1941), and *Recent Trends in Land Tenure in Texas* (Texas Agr. Expt. Sta. Bull. 641, 1944); J. H. Bondurant and W. C. Binkley, *Land-Tenure Classification and Areas in Kentucky* (Kentucky Agr. Expt. Sta. Bull. 421, 1942); G. G. Quackenbush and O. G. Lloyd, *Farm Tenure in Indiana by Type-of-Farming Areas* (Purdue Agr. Expt. Sta. Bull. 488, 1943); J. H. Southern, *Farm Tenancy Areas in Oklahoma* (U.S.D.A., 1944).

which deal with the legal aspects of landlord-tenant relations.[29] The procedure in each case was to have two men, usually a land economist and a person with legal training, interview various officials and legal personnel in respect to interpretation of the law as applied to landlord-tenant cases. The studies also invoked a search of state statutes and court decisions.

Although these bulletins are like the type of tenancy area studies in that they primarily serve to bring a body of information together, they do serve another purpose that fulfills a more definite research function. In these studies the investigators went out to ask people who would most likely know, what were the sources of legal doubt and confusion in respect to landlord-tenant relationships. The reports, therefore, flush out points of dispute. In this view, these reports on legal aspects can be regarded as important first steps in formulating problems for inquiry. None of them accomplish more than this so far as research is concerned. In the future it will be interesting to note whether, as has been often the case, these studies will now be regarded as finished, or whether the groundwork that has been laid will really be used as a cornerstone for further and more conclusive work.

Insofar as some of these reports include suggested changes in the law, such recommendations are not based on evidence in the bulletins to show the consequences in experience of the situation as it exists. This may actually mitigate against more thorough investigational work in the future because it leads to a false judgment that the preliminary probing represents a completed inquiry.

PLANTATION TENANCIES

The plantation system in parts of the southern United States presents a significantly different set of landlord-tenant problems from those found in most other parts of the country.

A study of the plantation economy was initiated by the federal Department of Agriculture in its early farm economics work. In 1916, a study was published which had as its purpose, not the resolution

[29] J. H. Dickerson, *Proposed Adjustments in the Farm Tenancy System in Missouri* (Missouri Agr. Expt. Sta. Res. Bull. 270, 1937); Marshall Harris, A. H. Cotton, and Rainer Schickele, *Farm Tenure in Iowa, V, Some Legal Aspects of Landlord-Tenant Relationships* (Iowa Agr. Expt. Sta. Bull. 371, 1938); H. R. Moore, *Some Legal Aspects of Landlord-Tenant Relationships in Ohio* (Ohio Agr. Expt. Sta. Rural Economics Mimeo. Bull. 119, 1939); L. J. Coleman and H. A. Hockley, *Legal Aspects of Landlord-Tenant Relationships in Oklahoma* (Oklahoma Agr. Expt. Sta. Bull. 241, 1940); H. W. Hannah and Joseph Ackerman, *Legal Aspects of Farm Tenancy in Illinois* (Illinois Agr. Expt. Sta. Bull. 465, 1940); H. A. Hockley and W. D. Nicholls, *Legal Aspects of Farm Tenancy in Kentucky* (Kentucky Agr. Expt. Sta. Bull. 418, 1941); H. A. Hockley and Harold How, *Farm Tenure Law in Kansas* (Kansas Agr. Expt. Sta. Bull. 303, 1942). Also see *Some Legal Aspects of Farm Tenancy in Oregon and Washington* (Northwest Regional Council, 1939).

of a problem, but an answer to questions as to what type of tenant receives the highest income and from what type of tenancy landlords make the highest return.[30] The question is answered on the basis of the averages and frequency distributions calculated for each of three groups of farms: those of share croppers, share renters, and cash renters.

In 1924 the Division of Land Economics published a comprehensive description of the plantation system, giving special attention to the labor supply and laborer and tenant relations on the plantations.[31] The study covers several hundred plantations throughout the cotton South. The problem is loosely formulated in terms of the "unusual shifting of agricultural workers on and off the farm," and "an exodus of laborers from the plantation" during the war. But the purpose is stated as only "To determine the nature of plantation problems and, as far as possible point out desirable methods of meeting difficulties."

The report gives an excellent and complete description of many aspects of the plantation system and raises a number of points which seem to be sources of difficulty, but the inquiry does not perform the function of pointing up these issues in terms of formulated problems or hypotheses for future research. Nor do the conclusions offer suggestions for improvement, but rather are limited to a brief statement of the advantages and disadvantages of the plantation system.

Although comprehensive reports have been published on the plantation system by land economists, rural sociologists, and farm management specialists, they are almost entirely descriptive.[32] In the latest comprehensive body of factual data brought together to describe this system, it is emphasized again that "No effort has been made to formulate a concrete working program. . . . The first important step . . . is that of objective analysis and careful appraisal . . ." and reasons presented for any programs of change are not to be considered in the near future.[33]

[30] E. A. Boeger and E. A. Goldenweiser, *A Study of the Tenant Systems of Farming in the Yazoo-Mississippi Delta* (U.S.D.A. Bull. 337, 1916).

[31] C. O. Brannen, *Relation of Land Tenure to Plantation Organization* (U.S.D.A. Bull. 1269, 1924).

[32] H. W. Blalock, *Plantation Operations of Landlords and Tenants in Arkansas* (Arkansas Agr. Expt. Sta. Bull. 339, 1938); E. L. Langsford and B. H. Thibodeaux, *Plantation Organization and Operation in the Yazoo-Mississippi Delta Area* (U.S.D.A. Tech. Bull. 682, 1939); J. G. McNeely and G. T. Barton, *Land Tenure in Arkansas II. Change in Labor Organization on Cotton Farms* (Arkansas Agr. Expt. Sta. Bull. 397, 1940); Harold Hoffsommer, *The Sugar Cane Farm* (Louisiana Agr. Expt. Sta. Bull. 320, 1940); Harold Hoffsommer, *The Resident Laborer on the Sugar Cane Farm* (Louisiana Agr. Expt. Sta. Bull. 334, 1941); J. G. McNeely, G. T. Barton, T. R. Hedges, *Land Tenure in Arkansas, III. Income and Changes in Tenure Status of Share Renters, Share Croppers, and Wage Laborers on Cotton Farms* (Arkansas Agr. Expt. Sta. Bull. 438, 1943).

[33] F. J. Welch, *The Plantation Land Tenure System in Mississippi* (Mississippi Agr. Expt. Sta. Bull. 385, 1943).

Land Tenure

It is rather surprising, in view of the historical tenacity of the ideal of ownership by the farm operator, that, at least in terms of numbers of published reports, a great deal more attention has been given to questions of landlord-tenant arrangements than to questions of the achievement of ownership. As has been seen, the earliest experiment station and federal department bulletins dealt with improvements in tenant leases.

OWNER AND TENANT COMPARISONS

In 1914 Missouri issued the first experiment station bulletin specifically devoted to a comparison of owner-operated and tenant-operated farms.[34] Whereas the 1910 Wisconsin bulletin and the 1911 federal bulletin had dealt only with tenanted farms, Warren's 1911 Tompkins County survey contained a few tables in which were compared the averages of rented and owned farms in respect to incomes, expenses, acreages, and yields of crops. The Missouri study follows this pattern. It was made "by means of the farm management survey" and consists mainly of parallel columns of averages and freqency distributions of data from farm management schedules for owner, part owner, and tenant farms.

The major points of comparison have remained fairly constant in many subsequent reports that have been issued over the years in Missouri and elsewhere. These are: income, system of farming, occupancy stability, and noneconomic social characteristics. (The Missouri study, however, has only scant information on one of the latter attributes — education.)

There is in the Missouri report no formulation of the problem or hypothesis to which the study is addressed, although at places in the report "evils" of "the present system of land tenure" are mentioned. For example: "The only reason why rural communities are as well off as they are at present, is because of the fairly large percentage of farms operated by the owners." But an increase in the percentage of tenancy is forecasted. In view of these problematic statements, and in view of the fact that the comparisons in the bulletin are between owners as a class and tenants as a class, it is surprising to note that the only recommendations for action are that the landlords should "rent their farms for a period of not less than five years" and should "compel the tenant to keep and feed livestock on the farm."

The weakness of these conclusions is obvious since nowhere in the

[34] O. R. Johnson and W. E. Foard, *Land Tenure* (Missouri Agr. Expt. Sta. Bull. 121, 1914).

report is there evidence that tenants who do meet these conditions also have good incomes, good farming systems, occupancy stability, or social advantages.

In several farm management studies during the decade following the Tompkins County survey, data for tenants and owners were compared as to farm incomes and farming system. These data were used in the important article on farm ownership and tenancy that Dr. Gray and his staff presented in the *1923 Yearbook of Agriculture*.[35] For comparison of stability of occupancy, they used agricultural census data. For comparisons of noneconomic social characteristics of owners and tenants, the data were taken from available rural social surveys and from a few current land economics investigations.[36]

One interesting comment is that up until this time, the Wisconsin-trained agricultural economists, Hibbard, Black, Lloyd, Grimes, and Falconer, had published studies that were restricted to investigation of farm leases within the tenancy framework. On the other hand, farm management survey work of men in the Warren and Spillman tradition provided some comparative data on owner and tenant operatorship.[37]

Beginning in 1922, Nebraska printed a series of reports of data collected in 1920 which provided quantitative comparisons of owner and part-owner farms and various types of rented farms. These reports presented such comparisons in respect to family composition, reading matter in farm homes, and participation in community life, as well as capital investments and farm equipment.[38] Except for such statements as "Farm tenancy and its effects are of interest and importance," there is no formulation of a tenure problem or hypothesis in any of these reports. Nor is there any apparent basis for the selection of data that vary from the comparative average size of the dooryards on

[35] L. C. Gray *et al.*, "Farm Ownership and Tenancy," *Yearbook, 1923* (U.S.D.A.), pp. 569–81, 589–96.

[36] Beginning about 1911, several church organizations and others interested in the rural country life movement initiated rural surveys in many counties over the country. A great deal of this work was done in Ohio, particularly by the Presbyterian church. The Inter-Church World Movement conducted surveys in more than 10 states. These surveys covered many aspects of rural life but emphasized particularly membership in social organizations, education, health, community stability, and the like. Most of them did not invoke a farm-to-farm canvass for information. It should also be noted that Warren's Tompkins County survey is offered as a "contribution to the country-life movement," but specifically leaves for "other kinds of surveys" such topics as "general social conditions of the communities."

[37] A partial exception to this is the work of the economics department at the University of Texas, done as a staff project, in which G. S. Wehrwein had a prominent part. See *Studies in Farm Tenancy in Texas* (University of Texas Bull. 21, 1915); and *The Land Problem in Texas* (University of Texas Bull. 39, 1915).

[38] J. O. Rankin, *Reading Matter in Nebraska Farm Homes* (Nebraska Agr. Expt. Sta. Bull. 180, 1922); *The Nebraska Farm Family: Some Land Tenure Phases* (Bull. 185, 1923); *Nebraska Farm Homes* (Bull. 191, 1923); *Nebraska Farm Tenancy: Some Community Phases* (Bull. 196, 1923).

share-cash and share-rented farms to the proportionate occurrence of accordions in owner and rented farm homes. Nor are there, of course, any conclusions in these bulletins in respect to action to be instituted.

The extent to which the use of standardized sets of deadly parallel columns can move research away from useful analysis is illustrated at many points in this series. In these bulletins the standard comparisons are those of all owner and all tenant farmers even though the data were taken from widely separated parts of the state. Consequently, in several instances average differences between these groups have to be explained on the grounds that certain tenure forms predominate in different parts of the state so that the apparent tenure differences are in fact areal differences. In Bulletin 196, for example, is this statement: "Share tenants exceed even part-owners in average distance to community activities, because they are also concentrated largely in the westernmost areas but do not have the compensating influence of being concentrated in the Walthill area also as the part-owners are." This example of meaningless construction and dissection of averages is the obverse side of those mentioned in reviews of certain land utilization studies. In the latter data were presented in the form of standardized county or district averages even when it was not clear that these areas had any analytical significance.

A few years after the Nebraska reports were issued, North Carolina issued two bulletins in which data collected in a sociological survey of living standards are presented on an owner and tenant comparison basis. These studies represent the height of the statistical mania as it affected tenure studies. The first of these reports is presented merely as "a statistical comparison of living conditions among the white owner and tenant families of the same area." [39] Like the Nebraska bulletins, this one has no stated problem and no hypothesis, and no recommendations are made. The chief difference is that besides giving frequency distributions and mean figures for the various items as between the owner sample and the tenant sample, the North Carolina report also presents the modal and median figures for some items. According to a statement in the conclusions, the "main thesis of the study is that farm family living conditions are the function of a number of interacting factors and that adequate analysis of these varying conditions . . . must be based upon the measurement of the relative influence of these factors." But there is nothing else in the conclusions to indicate the possible usefulness of the study, or what is accomplished in an adequate analysis, if this report is to be so characterized.

[39] W. A. Anderson, *Farm Family Living Among White Owner and Tenant Operators in Wake County, 1926* (North Carolina Agr. Expt. Sta. Bull. 269, 1929).

In a companion bulletin, an even more refined attempt to measure relative influences of factors is made.[40] This report represents the extent to which some social scientists went in the late twenties in applying formal statistical correlations. It is a long report based on a series of tables showing the gross, partial, or multiple correlation among various data from tenant schedules as compared with owner schedules. The purpose "was to investigate the changes in family living conditions in the two groups as changes occur in some of the factors effecting [sic] these conditions." The hypothesis is that "no adequate knowledge will be obtained of family living conditions . . . until it is recognized that family living is the product of many factors which affect each other [and] that . . . these significant factors must be discovered and their influence measured."

Since seven of the thirteen items used in the analysis are the percentage of the total family expenditures that go to each of these seven things, it is not hard of course to establish that these factors "affect each other." In the conclusions, it is pointed out that "budgetary influences" (that is, the income—and percentage of income—spent for various items) is "largely responsible" for changes in the percentage spent on different items.

While this report is primarily a study of living standards rather than a tenure study, it is mentioned here not only because it is structured on owner-tenant comparisons but also because it shows that although the sociologists may have centered more attention on such matters as living conditions and social contacts than would otherwise have obtained, in this instance at least there is no indication of any special interest in, or awareness of, social problems. On the contrary, the work is even more descriptive and less problem oriented than most other tenant-owner reports. It goes further than any other bulletin dealing with owners and tenants in the direction of manipulating mass data of mathematical relationships instead of handling evidence in such a way as to expose the internal patterns of experience in the lives of the people in the area studied.

Further indication of the interest of sociologists in owner and tenant comparisons about this same time is an Oklahoma study of the relation between owner, part owner, tenant, and cropper status and various aspects of their social behavior, and a special investigation of the extent and costs of farm moving.[41] In a South Dakota study, com-

[40] W. A. Anderson, *Factors Influencing Living Conditions of White Owner and Tenant Farmer in Wake County, 1926* (North Carolina Agr. Expt. Sta. Tech. Bull. 37, 1930).

[41] O. D. Duncan and J. T. Sanders, *A Study of Certain Economic Factors in Relation to Social Life Among Oklahoma Cotton Farmers* (Oklahoma Agr. Expt. Sta. Bull. 211,

parisons of a few items such as age and care of machinery and church membership are made between tenure groups.[42] But the recognized tenure groups are constructed, not on the basis of ownership and tenant status, but on the basis of "freedom of control over management" and "security in possession of such rights." Under this procedure, for example, "long-term debt-free tenants" are in the same group as the full owner with "light debts."

In 1932 Delaware published a land tenure bulletin of considerable length containing a great deal of factual information about land tenure in four areas scattered over the state.[43] Although parts of this report are devoted to the question of lease arrangements on rented farms, the bulletin also gives a good share of attention to owner-tenant comparisons.

In the description of the problematic situation, two factors are emphasized: (1) farms have passed to landlords instead of to operating owners and (2) outmoded rental contracts are causing "numerous conflicting interests." The objectives of the study are to determine or to indicate the factors affecting the trend of tenancy and ownership and the sources of landlord-tenant conflict, yet the only purpose pointing toward a possible solution is "to outline a rental contract" that will "promote more harmonious relations."

The structure of the bulletin consists in the main (1) of graphic and tabular comparisons of average owner and tenant farm figures for each of the sample areas and (2) of landlord and tenant figures for each area and for other groupings.

In this study there is more than the usual flexibility in the regrouping of the data and the data are very comprehensive. In spite of these advantages, however, the report is still an exemplar of the widespread research practice of dealing on an item-by-item basis and of limiting the analysis of each item to a gross comparison of relative magnitudes, rather than of holding together, in terms of their actual occurrence, series or chains of connected items.

This technique is well illustrated, for example, by two consecutive tables which show that for one area about three fourths of the landlords feel that their present leases are fair and that in the same area about three fourths of the tenants regard their leases as fair. This separation makes it of course completely impossible to follow what

1933); J. T. Sanders, *The Economic and Social Aspects of Mobility of Oklahoma Farmers* (Oklahoma Agr. Expt. Sta. Bull. 195, 1929).

[42] W. L. Slocum, *The Influence of Tenure Status Upon Rural Life in Eastern South Dakota* (South Dakota Agr. Expt. Sta. Circular 39, 1942).

[43] R. O. Bausman, *Farm Tenancy in Delaware* (Delaware Agr. Expt. Sta. Bull. 178, 1932).

would seem to be an important lead: Are the satisfied tenants the same ones whose landlords are also satisfied? In the same area a comparatively high percentage of tenants indicated no intention of becoming farm buyers. The obvious question, were the study more analytical, would be: Is it on the farms where tenants are satisfied with their contracts that the landlords regard arrangements as satisfactory and the tenants do not intend to buy? In other words, the explanation of the problem lies in the interaction of factors within each landlord-tenant relation and in the sequence of each relation over time. The technique of gross numerical comparisons on an item-by-item basis can at best throw up suggestions of possible internal relations; the actual relations within which explanations lie must be exposed as existing in each case.

Finally, so far as this Delaware report is concerned, it is not surprising that the last paragraphs of the bulletin report that although the "study has pointed out many weaknesses in the present rental contract," . . . "it is probable that generally, and at least for the present generation, the present rental contract is about the most workable contract under existing Delaware conditions." And all the scores of tables of data on owner-tenant comparisons lead to no recommendations — merely the prediction that tenancy will decrease because of "a decrease in the part played by inheritance in farm ownership," "a decrease in sentiment in farm ownership," and a "continuing scarcity of good tenants." These factors are mentioned in the introduction to the report, but nowhere in the bulletin is there any evidential proof of the importance of these enumerated causal factors.

Following the revival of public interest in land tenure, instigated by the report of the special President's Committee on Farm Tenancy in 1937, there was a surge in experiment station projects relating to farm tenure. Several states published bulletins in which census data and other readily available data were used "to indicate the distribution and growth of farm tenancy . . . and to describe some of its economic and social characteristics."[44]

In the South owner-tenant studies were issued in Mississippi, Georgia, and Kentucky.[45] The Mississippi report, a brief one, is inter-

[44] J. H. Southern, *Farm Tenancy in Oklahoma* (Oklahoma Agr. Expt. Sta. Bull. 239, 1939). Also see J. A. Baker and J. G. McNeely, *Land Tenure in Arkansas, I. The Farm Tenancy Situation* (Arkansas Agr. Expt. Sta. Bull. 384, 1940); Rainer Schickele, *Farm Tenure in Iowa, II. Facts on the Farm Tenure Situation* (Iowa Agr. Expt. Sta. Bull. 356, 1937).

[45] Dorothy Dickins, *Owner Farm Families in Poor Agricultural Areas and Cropper Families in Rich Agricultural Areas* (Mississippi Agr. Expt. Sta. Bull. 359, 1941); W. T. Fullilive, J. C. Elrod, and W. E. Hendrix, *A Study of Farming by Tenure of Farms in Terrell County, Georgia* (Georgia Agr. Expt. Sta. Bull. 234, 1944); J. H. Bondurant,

esting because it is restricted to a partial answer to a very specific action question: Is "farm ownership on poorer areas" a better alternative than "share cropping in rich areas"? Put another way the hypothesis reads: If white croppers were moved from highly productive land and were established as owner-operators on poorer land, then they would enjoy better family living. The data used in the analysis consist of average and frequency distribution comparisons of income and family expense items of a group of owner farmers in a very poor part of the state and of croppers in another area physically suitable for agriculture.

That the data and analysis presented cannot ground the hypothesis is recognized in the bulletin. Nevertheless, the hypothesis is so defined that the partial modification of it suggested by the inquiry represents a modest but clear-cut contribution to the problem under consideration.

The Georgia report consists of a statistical comparison of average and percentage figures for different tenure groups in respect to farm incomes and systems of farming. While the problematic situation is described as that of doubt as to whether ownership is a "solution to all tenure problems," the objective of the study is "to examine the characteristics of the farming businesses under the major tenancy systems in a county" and to compare them in some factors "generally associated with a successful and well-balanced farming economy." But in this report the croppers are not regarded as a tenure group, and therefore only incidental reference is made to the very class which has aroused more public concern than any other tenure group in American agriculture. In the Kentucky report, on the other hand, "special emphasis was placed on the problems of the cropper."

Both the Georgia and the Kentucky reports illustrate a persistent difficulty caused by using techniques developed in one field of study for analysis of questions in a very different area. In farm management, special accounting terms have been devised specifically to eliminate the influence of the farmers' real estate and property status from analyses of operating efficiency. In the calculation of "labor income," for example, 5 per cent of the total value of the farm is charged as an expense item whether the operator actually owns the farm (in which case his reported available income is thus drastically cut) or whether he actually is in debt for the farm (in which case he may not only have to pay this charge but may also have to meet payments on principal besides).

Land Tenure in Southern Logan County, Kentucky (Kentucky Agr. Expt. Sta. Bull. 464, 1944).

It is clear, therefore, that judgments as to the economic position of owners and tenants are severely twisted by use of such figures, both by understating the actual income of owners with some net worth and by misstating the position of heavily indebted persons. In the Georgia study, for example, the relative average positions of a group of owners and tenants are reversed when interest of $186 is deducted from a farm income of $574 for the owners and only $37 is deducted from a farm income of $508 for the tenants. In the Kentucky study the average income of owner-operators who hire share croppers turns out to be within only about fifteen dollars of the average incomes of the croppers themselves. But it is presented only on a "labor income" basis.

Other recent studies of farm tenure comparisons have been published in Minnesota and Nebraska.[46] Although the Minnesota report is primarily a rental lease analysis, it contains more information than is found in the bulletins listed above in that category. This study, like most general tenure survey bulletins, is designed "to present . . . data on the development of farm tenancy . . . and to portray a factual picture of its present status and the factors that contribute to this situation." In terms of inquiry to resolve problems, however, the study is also described as serving as a preliminary for subsequent studies designed to test present practices and to devise improvement in these practices. The short summary of problems section near the end of the report is therefore more important in terms of the progress of scientific inquiry in the solution of problems than the longer summary chapter of factual information.

Somewhat in contrast, the Nebraska study includes in its stated objectives, the calculation of adjustments "to improve present conditions"; and a closing section on "Possible Adjustments" is presented. This bulletin is another illustration of failure to arrange materials in the report so as to constitute evidence that is relevant to the conclusions. For example, the recommendations stress the desirability of security of tenure. But in spite of the fact that 93 per cent of the tenants expressed a desire for leases longer than one year, the report shies from this idea in favor of specified notice dates for lease termination. In the data presented, it is clear that some of the tenants already have leases longer than one year, and some already have definite termination notice dates. But nowhere in the report are the available data on cropping systems, productivity, valuations, erosion, living

[46] G. A. Pond, *Farm Tenancy in Minnesota* (Minnesota Agr. Expt. Sta. Bull. 353, 1941); G. H. Lambrecht and L. W. Wallin, *Farm Tenancy in Box Butte County, Nebraska* (Nebraska Agr. Expt. Sta. Bull. 336, 1942). Also see Rainer Schickele, *Farm Tenure in Iowa, IV. Farm Tenure Conditions in Palo Alto County* (Iowa Agr. Expt. Sta. Bull. 364, 1937).

conditions, social life, income, or mobility analyzed on the basis of those instances in which the leases are long or short or in which the termination notice data is specified or unspecified. In other words, the possible adjustments are not scientific conclusions, even though the data collected could have been arranged as testing evidence had these adjustments been framed as a beginning hypothesis and had the hypothesis then been used to direct the sifting and ordering of facts as evidence.

By way of recapitulation some general comments may be made in respect to these owner-tenant comparison reports. In the first place, it may be noted that some of them are essentially lease analysis bulletins with some additional information on owned farms that is compared with similar data for rented farms. Secondly, it is recalled that the owner-tenant farm comparisons are based on what may be summarized as (1) income, (2) farming systems, (3) stability of occupancy, and (4) family living and social factors. These studies have been made by men whose other work has been in farm management, rural sociology, or land economics. In all cases, however, the standard technique has been to show the relation between operator-ownership and tenancy by comparing gross quantitative data from both.

It is clear, whether stated specifically or not, that the assumption of these studies is that high incomes, nonexploitive farming systems, stable occupancy, high living standards, and social participation are desired outcomes, and that tenure status is strategic in the process of attaining these postulated consequences. But in very few reports is there any definite conclusion or suggestion as to what action in respect to tenure status can be instituted that will be consonant with the desired consequences; and even in the reports that make suggestions, there is no evidence warranting an assertion that if such and such actions are instituted under the stated conditions, the stated outcomes are experienced.

Mention should be made at this point of the fact that some studies dealing specifically with the consequences to soil conservation of ownership or tenancy will be treated separately in Chapter VIII. However, reference to one of these studies emphasizes the foregoing recapitulation. In 1939 the federal Department of Agriculture issued a study based on a special tabulation of census data on corn belt farms, designed "to determine whether or not significant differences in land-use patterns and livestock enterprises exist between owner-operated farms and tenant-operated farms in the Corn Belt."[47] In

[47] J. A. Baker, *Tenure Status and Land Use Patterns in the Corn Belt* (U.S.D.A. Land Economics Report 5, 1939).

some 30 tables, over 5000 owner farms and 7000 tenant farms are divided into eight areas and are then described by frequency distributions and averages of land use and livestock enterprise items. In the conclusion of this report it is, however, carefully stated that the study is "merely . . . exploratory" and that it contains "a strong implication" that tenure status causes soil exploitation. But the real function of the study is to "form the basis for more detailed and comprehensive investigations of the field-survey type" to provide "factual data" for "an objective evaluation of . . . tenancy with reference to one of the most pressing land-use problems . . . soil erosion."

For the purposes of this study the foregoing quotation is important, not only because it correctly emphasizes the limitation of quantitative comparisons of large groups of data to suggestions for further analysis, but also because it indicates that exactly 25 years after the first field study comparing owners and tenants in which stress was placed on exploitive farming practices, an analysis of data on over 12,000 farms can still only suggest that more field studies should be undertaken.

Two field studies of land owners and tenants on irrigation projects consist mainly of comparisons of data on farming systems and further illustrate the fallibility of customary procedures.[48]

The concluding remarks specifically recommend "some organized efforts to encourage the purchase of land by worthy tenants" and state that "quite definitely . . . ultimate ownership is the desirable type of tenure." This last conclusion from the Montana bulletin is based on a series of bar charts indicating that owners have somewhat more animal units per farm, somewhat more soil-conserving crops, and the like. But clearly, the objective of any program to assist tenants to become owners would be more than to shift the percentage of crop income from, say, 64 to 59 per cent, or to increase farm value per acre from $73 to $87. And merely to quote these pieces of evidence in terms of the conclusions with respect to action is to reveal the limitations of this type of research analysis so far as recommendations for action to be instituted are concerned.

In the Idaho study the same procedures for arranging the data are further developed so as to divide the owner and tenant farms by soil type. In this study are a great many apparently significant comments, which suggest awareness of important sources of difficulties. But these are only vaguely related to the manner in which the data

[48] P. A. Eke and H. F. Brown, *Influence of Tenancy on Types of Farming and Agricultural Income by Soil Types, Minidoka Irrigation Project* (Idaho Agr. Expt. Sta. Bull. 222, 1937); P. L. Slagsvold, *Land Ownership and Tenure, Huntley Irrigation Project* (Montana Agr. Expt. Sta. Bull. 385, 1941).

are presented so that their evidential force is diminished. For example, the "concluding statement" differentiates "two general types" of tenant farms for which different recommendations are made; but nowhere in the report are the data presented in terms of these two types. In the conclusions, these two are described in terms of soil quality, the character of ownership, the history of past ownership and operation, and location in respect to cities. Although apparently these strategically different types are clearly in existence, the data presented barely suggest their presence, for the evidence consists only of averages or percentages of all tenant farms lumped together or divided on the basis of soil type, kind of crop, or size of farm.

TENURE PROCESSES

It would be accurate to characterize almost all the material in the many land tenure bulletins already referred to as static. The data are such as to present a cross-section of certain tenure elements at a given moment of time. Whether the data describe crop yields, provisions of leases, farm income, assets, membership in organizations, age of operator, or years of occupancy, they show the current status of these items. They do not show developments over time.

Public interest in land tenure questions, however, can hardly be separated from the concept of change, development, or process. When the nation was shocked at the revelation of 25 per cent tenancy in 1880, the pertinent questions were: (1) How did this come about and will it increase in the future, and (2) can those who are not now farm owner-operators attain that status?[49]

Some of the early articles on farm tenancy referred to in Chapter II traced the historical, statistical trend in the growth of farm tenancy and gave conceptual interpretations and explanations; but early empirical research dealt with existing lease conditions and static comparisons of farm data for owner and tenant farms.

About 1914 Spillman in the U.S. Department of Agriculture had become interested in the process by which individuals actually attained ownership of land. In 1917 he published the results of a study in which more than 2000 farm owners were classed on the basis of the type of tenure (son at home, hired man, tenant, owner) they had each experienced. One fifth of the cases were owners who had actually spent some time in each status, and the range of these experi-

[49] Cf. W. B. Bizzell, *Farm Tenantry in the United States* (Texas Agr. Expt. Sta. Bull. 278, 1921), pp. 110–11, sections on "Increase in Farm Tenantry and Its Cause" and "Tenantry, a Transition in Farm Operations."

ences was referred to as the agricultural ladder.[50] In the 1920 census, data were obtained to throw more light on this process, and a special monograph was prepared on the length of time that owners and tenants spent in the various tenure stages. Because these data were grouped by the period in which owner status was reached, and because account was not taken of the effects of mortality, occupational shifts, and retirement, the results erroneously suggested that over time the periods of nonownership status were lengthening.[51] Nevertheless, the work of Spillman and the census article represent a first attack on questions of tenure experience over time.

A second way of looking at the dynamic aspects of farm tenure is to view the process of change in the holding of land in a given area over time. The first study of this type was issued in Wisconsin in 1919.[52] On the recommendation of an American Sociological Society committee that field studies of "the social aspects of tenancy" emphasize "the shifting of farm tenants," an investigator obtained a 10-year occupancy history of 500 farms in four Wisconsin townships.

With no further developed problem formulation than this desire to obtain a certain type of data, it is not surprising that "the main statistical facts of the study are presented in table form, without, however, any attempt at this time to interpret them," or that the bulletin has no conclusions or recommendations.

The report is nonetheless of methodological interest because it represents an attempt to handle data on the sequence of events on 500 different farms over a 10-year period. The procedures used are of several types, but they are not equally successful in respect to keeping sight of the actual processes of successive occupancies and tenures on the farms. One procedure is to add for each year the occurrence of various types of shifts in occupancy or holding. When this is done, most of the accompanying text is devoted to an interpretation of the table, for the sequence of the cases is hopelessly lost. A second procedure is to summarize each year's shifts into a sufficient number of groups to reveal "the story, year by year, of how many of the original farms" have shifted in forms of occupancy or tenure. This type of table can better express the processes of change on individual farms

[50] W. J. Spillman, "The Agricultural Ladder," *American Economic Review* (Supp., March 1919).

[51] E. A. Goldenweiser and L. E. Truesdell, *Farm Tenancy in the United States, 1920* (U.S. Census Mono. IV, 1920). Cf. Gray *et al.*, "Farm Ownership and Tenancy," *Yearbook, 1923*, pp. 556–61; Black, ed., *Research in Agricultural Land Tenure*, pp. 7–8; and G. H. Von Tungeln, "Some Observations on the So-called Agricultural Ladder," *Journal of Farm Economics*, vol. IX, 1 (Jan. 1927).

[52] C. J. Galpin and E. F. Hoag, *Farm Tenancy, An Analysis of the Occupancy of 500 Farms* (Wisconsin Agr. Expt. Sta. Res. Bull. 44, 1919).

than the first, but it is still not entirely free from the danger of misinterpretation. A third technique used is that of segregating a significant and well-defined subgroup and presenting a table in which is given the relevant information on each case separately by years. A fourth technique is to summarize the groups of cases in outline form and describe them by statements in subseries, as for example: (I) Owner-operator retired, (A.) still owning farm, (1.) and living on own farm, (a.) with son-tenant. Under these last two procedures, the individual experiences are preserved so that they are clearly seen even though ordered into groups.

Attention to the experience both of an area and of individuals is given in a third study of land tenure processes published in 1922 by the Division of Land Economics.[53] The early pages of this report give as clear-cut an indication of an existing problematic situation as any that has been reviewed. "The region is of special interest . . . because of the social and political unrest arising from its tenure problems, which at times has been a major factor in politics in the State and attracted national attention." After 1900 a landlords' "bonus system aroused violent antagonism," a renters' union's demands led to a federal investigation and passage of a state anti-bonus law. "Evidently . . . conditions in the black land have been more conducive to a rapid increase in tenantry than [elsewhere], and it is the discussion of this growth, its causes, and its effects, that has aroused nation-wide interest in the land problem of this area." After this convincing statement of the problematic situation, however, there is no suggestion of a formulation of this problem nor of a general working hypothesis. On the contrary, the project centers on analysis of available census data and "mainly on data . . . from 368 farm operators."

The analysis of the causes of the rapid rise in tenancy in the areas is based wholly on over-all census and tax record data which show that with a change from livestock farming to crop farming, large grazing units were split up into many smaller rented units at a time of heavy immigration into the area, and that since that time land values had risen with the rapid rise in the price level. Actually, of course, these data indicate some of the conditions under which tenancy increased, but they neither show how the increase took place nor why it occurred the way it did.

The second section of the report is comparable to the studies of

[53] J. T. Sanders, *Farm Ownership and Tenancy in the Black Prairie of Texas* (U.S.D.A. Bull. 1068, 1922). See W. S. Scarborough, *Tenancy and Ownership Among Negro Farmers in Southampton County, Virginia* (U.S.D.A. Bull. 1404, 1926), which has a very similar pattern of analysis.

leases, rented farm income, and owner-tenant comparisons previously reviewed in this chapter. Two points should be noted in this analysis, however.

In the discussion of lease contents a specific conclusion is given: "If legal provision were made for compensation of tenants for improvements put on with the landlord's consent, the present problem of tenant housing would be less acute." This statement clearly fits the interpretation of hypothesis as an "if-then" suggestion of action and consequences. But is it a beginning hypothesis or a conclusion? In fact, it is neither. Although it is offered in the spirit of a conclusion, there is nothing in the entire bulletin to warrant it. It is stated that "Thirty-six farms were found where the landlord had agreed" to this provisions already, but there is no attempt to show that where this action had already been instituted an improvement in housing resulted. The report merely goes on to say that landlords "complain that they cannot" improve housing "because their returns from the farm do not justify such expenditures"; yet elsewhere the data show an average landlord's return of 5.9 per cent on capital plus a land value increment equal to a "net compound annual interest of from 8 to 9 per cent."

It was previously mentioned that the standard labor income figure used in analyses of farm management efficiency is not suited to tenure status studies. This point is recognized in the 1922 bulletin, for the income analysis leads up to a comparison of "actual disposable net income" of the farm operator, "in which interest and wages are deducted only when actually paid."

A section in U.S.D.A. Bulletin 1068 on "domestic, social, and educational conditions in relation to tenure" does not differ significantly from the Nebraska study and other studies previously mentioned. The special analysis of this report is the tenure and financial "agricultural history of farm operators." In the analysis of "tenure stages" there is an important illustration of the difference between aggregate and individual sequence data. In his early work Spillman classified the farmers according to the pattern of experience each had had. Thus, all those classified as FTO (F, laborer on home farm; T, tenant; O, owner), for example, had actually passed through each of these stages. In U.S.D.A. Bulletin 1068, however, there is a table showing the percentage of all operators who have been in various stages. The table shows that 35 per cent have experienced operator-ownership; and it also shows that 33 per cent have experienced nonfarm employment. It is clear that if, for example, these percentages referred to an essentially identical group among those surveyed, the implications would be greatly different than if this 33 per cent was included in the 44 per cent who had experienced cropping, and if, say, none of the group had attained ownership. In other words, with the sequence of experience within each case destroyed, it is possible only to surmise what actually occurred; with the sequence held intact, this information is clearly available.

Furthermore, when the report turns to the financial history of farm operators, the analysis is thrown further away from the sequence of tenure stages and is structured on a subdivision of each of the three main tenure groups — those whose average annual accumulations from earnings have been high, medium, or low. Comparisons in averages and percentages of these subgroups for numerous items are made. The conclusions of this analysis are of course in the form that "high accumulators" average more or less in respect to a given item than do "low accumulators." The limitations of this type of analysis may be seen by reference to the summary, which reports that those "who accumulated wealth most rapidly included those" who had worked most consistently as farmers, who had more diversified farming, who raised more of their family food supply, and who moved less frequently. The chief question to be asked is whether these findings even imply, let alone recommend, that tenants and croppers would be better off if they did not move off their farms from time to time, if they did not accept nonfarm work on occasion, and if they cut down their cotton acreage — if they could. If the findings do not mean this as even an implied suggestion, then they mean that the situation as it is is understandable, but that the sharply defined problematic situation described at the outset remains no less problematic because a social science inquiry has been conducted in respect to it.

In 1923 a social survey of Cedar County, Iowa, was published in which the whole emphasis is on the comparison of owner and tenant farms.[54] The data include various aspects of family composition, social participation, and education; but chief attention centers in the progress on the agricultural ladder. Also, some data are arranged to show the changes in occupancy of the farms of the area. In keeping with the stated purpose, these data comprise an "inventory of . . . present conditions," and the report passes rapidly from table to table and topic to topic but with no direction and no conclusions.

Agricultural ladder studies were reported in Wisconsin and Ne-

[54] G. H. Von Tungeln, E. L. Kirkpatrick, C. R. Hoffer, and J. F. Thaden, *The Social Aspects of Rural Life and Farm Tenantry in Cedar County, Iowa* (Iowa Agr. Expt. Sta. Bull. 217, 1923).

braska in 1926 and 1928.[55] Both are aimed at the "apprehension that young farmers are unable to buy farms as they formerly did." The Wisconsin study emphasizes geographic differences in the types of tenure sequences into which the operators are grouped, but both include descriptions of the various stages of tenure in terms of sizes of farms operated and length of time spent in each stage. The Wisconsin analysis points only to the conclusion that "ownership though somewhat slower of attainment is still the rule." In neither study are there any concluding suggestions whether the attainment of ownership could be expedited or how it might be done.

One of the largest studies of tenure processes was completed in North Carolina in 1937.[56] This bulletin reports on a study of nearly 3500 farms in six different areas of the state. The report, consisting of 180 pages of fine print, presents a massive array of quantitative data in 79 tables and 59 figures. The "general purpose is to determine the effect of depression, agricultural adjustment, and business recovery upon some of the basic social and economic processes in rural North Carolina." Since depression, adjustment, and recovery are merely another way of saying recent years, there is no specific action element in the situation to which the survey is directed; and the so-called basic social and economic outcomes are mainly tenure status but also marriage rate, birth rate, population characteristics, and the like. The nearest the report comes to being specific on these points is in a statement in the final chapter that "It is firmly believed that landlord-tenant relations in the long run will be greatly improved by the Agricultural Adjustment Program." Because the "study showed . . . that there is, even in 'normal' times, a constant movement up and down the agricultural ladder" and because "there were just about twice as many movements up as down the ladder" in 1934 and 1935, "it is more or less absurd to attribute all of the displacements of 1933, 1934 and 1935 to the Agricultural Adjustment Program." Had this issue been posed as the question of the study, it would certainly appear that the answer could have been obtained more expediently, more directly, and more convincingly.

A report on land ownership and tenure in Imperial Valley, California, deserves comment because it is based entirely on (1) existing data on land use and tenure from the irrigation district office, the U.S. census, the county tax assessor, published soils surveys and other

[55] B. H. Hibbard and G. A. Peterson, *How Wisconsin Farmers Become Farm Owners* (Wisconsin Agr. Expt. Sta. Bull. 402, 1928); J. O. Rankin, *Steps to Nebraska Farm Ownership* (Nebraska Agr. Expt. Sta. Bull. 210, 1926).

[56] C. H. Hamilton, *Recent Changes in the Social and Economic Status of Farm Families in North Carolina* (North Carolina Agr. Expt. Sta. Bull. 309, 1937).

experiment station reports, and (2) the personal knowledge and qualitative observations of the investigator.[57] The results are informational; they do not include any recommendations. But they do clarify and pose specific problems for the solution of which other studies should be made. The danger is that the multicolored maps and impressive tables of data, composed from already available data, may give such an impression of exhaustiveness as to screen the fact that the report only really poses and sharpens issues and in no sense arrives at or confirms recommendations for solution.

Three studies have recently been published all dealing with the issue as to whether the operator-ownership ideal can be attained, and what action needs to be instituted if it is to be more nearly reached. An Oklahoma study raises this issue and "analyzes a large collection of factual data relating to farm ownership in Oklahoma, for the purpose of finding out what . . . accelerate[s] the rise to ownership . . . [or] hinder[s] it."[58] The factual data are arranged to provide gross quantitative comparisons between owners, tenants, and other farm workers; and the analysis runs in terms of these magnitudes. Thus, for example, although the data are not regarded as self-sufficient or conclusive, the "percentage of owners is smaller among families with more than six or seven children than among those with fewer," and such evidence seems to point to "need of stabilizing the size of the family at some point in the neighborhood of five children for movement toward farm ownership to be the least hindered." Fortunately, however, the study, in its summary does not emphasize such unfounded conclusions—unfounded precisely because the action-consequence is not revealed in any evidence. Instead, the generalization is that "the study shows that farmers will need assistance at the hands of the public, particularly in the form of enlarged credit facilities, if farm ownership is to increase relatively in the future." But unfortunately, this too, is unwarranted as a scientific conclusion of the study, for nowhere is there any evidence as to the effect of credit on attainment of ownership.

In contrast are two Wisconsin studies of ownership attainment. One of these investigates the tenure process in Lafayette County where tenancy is the highest in the state,[59] and the other deals with an area

[57] Adon Poli, *Land Ownership and Operating Tenure in Imperial Valley, California* (U.S.D.A., 1942). Cf. Philip Greisinger and G. W. Barr, *Agricultural Land Ownership and Operating Tenures in Casa Grande Valley* (Arizona Agr. Expt. Sta. Bull. 175, 1941).

[58] R. T. McMillan and O. D. Duncan, *Social Factors of Farm Ownership in Oklahoma* (Oklahoma Agr. Expt. Sta. Bull. B-289, 1945).

[59] L. A. Salter, Jr., *Land Tenure in Process* (Wisconsin Agr. Expt. Sta. Res. Bull. 146, 1943).

in eastern Wisconsin where practically all farm operators are owners.[60] In both inquiries the problem situation involved perplexity in respect to failure to attain ownership; in both, the problem was formulated in terms of the loss of ownership between generations; and in both, the hypothesis was that if practices were invoked in the transfer of farms between generations, more ownership and less tenancy would result.

The chief characteristic of the procedure used in these studies is that the farms to which the data apply are held as recognizable units, throughout the Lafayette County analysis and in most of the eastern Wisconsin study. For example, it is pointed out that lands are transferred by various designated routes and the elements which make these routes different are made known. Furthermore, it is shown that some owners and occupiers of land stand at different points in the same routes along which other farms and other farmers have traveled.

In the Lafayette report, for instance, nearly all debt-free farms were found to have been transferred as debt-free farms. Cases that are exceptions to this rule are individually explained. Similarly, in the eastern Wisconsin study nearly all farms on which debts have been reduced are shown to have been transferred under special intrafamily practices; and exceptions are individually explained.

These studies use a research procedure in which enough cases are included to cover a sufficient variety of experiences to check practices with consequences, and the cases are grouped in terms of their internal pattern or sequence of experience. This procedure offers a means of avoiding the limitations of other techniques or at least of providing one in which relations suggested by gross quantitative relationships can be tested against their actual interaction in experience.

It should be pointed out that since the Lafayette study was formulated in terms of full ownership as the end-in-view and since very few cases were found in which purchase debts had been cleared, there was little opportunity to expose practices that could be termed successful. Therefore, of the four specific recommendations in the concluding section, only one is based on observed practice in experience and is actually tested; the others have scientific status only as suggestions. In the eastern Wisconsin study, on the contrary, cases of failure of consequences (tenancy) were relatively scarce; and the study was formulated in terms of progress in equity accumulation. Therefore, in this instance, each of the four specific recommendations has a higher degree of warrantability.

[60] K. H. Parsons and E. O. Waples, *Keeping the Farm in the Family* (Wisconsin Agr. Expt. Sta. Res. Bull. 157, 1945).

Here it should be noted that in these studies the selected areas are at the extremes in terms of the existence in Wisconsin of the problem under consideration. They were picked as such in order to provide the strongest tests of the phenomenon over which there was doubt. In terms, then, of giving a representative description of what is happening, the Lafayette study is more useful because the extreme of the existence of tenancy in Wisconsin is more like other parts of the United States than is the other extreme of the absence of tenancy in Wisconsin. Also, therefore, the Lafayette County report is probably more important to understanding why tenancy conditions generally are as they are. But in terms of importance for posing alternative means of action which, if instituted, would retard the growth of tenancy, the eastern Wisconsin study is essential, for it is in such an unrepresentative area (practically by definition it is in such an area) that going experiments of how to attain postulated outcomes can be found by the social scientist.

Brief mention should be made of a small mimeographed report issued in New York dealing with the ownership and mortgage history of one town.[61] This study is not formally set in terms of any problematic situation, formulated problem, or hypothesis; nevertheless the entire analysis runs in terms of comparing the ownership history of farms on land classified in the New York land classification system as III and IV against those on land classes V and VI, the better lands. Because of this mold into which the data are cast, the conclusions are also mere comparisons, such as that a higher percentage of farms on the poorer lands have never been out of debt, have changed hands more frequently, and so on. If, however, the purpose of this inquiry were to analyze success and failure in attaining debt-free ownership, clearly the emphasis should be on the different patterns of experience on those places that were successful as contrasted with those that were not, so long as even the most elementary analysis at once reveals that some of each experience group are on good and poor lands.

The failure to rearrange the facts in this report so they constitute testing evidence is made particularly clear because on two pages an interesting device is constructed and used. The historical pattern of ownership experience of each of 24 farms over a century is symbolically revealed by the device in such a way that comparisons and contrasts of these chains or sequential patterns of experience are exposed. The possible usefulness of this technique for handling sequen-

[61] H. F. Degraff, *The Ownership and Mortgage History of Farms in the Town of Newfane, Niagara County, New York* (Cornell University Department of Agricultural Economics AE 341, 1941).

tial evidence is lost, however, because of the dominant procedure of casting the data into comparisons of gross magnitudes between apparently almost irrelevant land classes, and because of the lack of stated purposes for the analysis.

A word must be also given in reference to a land tenure process study made outside the experiment station and federal Department of Agriculture agencies. In 1941 the University of Chicago published a report on a Nebraska community, made by the great grandson of the founder of the community.[62] In its tracing of the tenure process, this study has some of the characteristics of the two Wisconsin bulletins previously mentioned, in that there is some attempt to group the farms of the area by subseries patterns of the characteristics of their tenure sequence. The study differs in that the Wisconsin studies deal with a problem that is formulated with operator-ownership as the postulated outcome and are thus restricted to the analysis of why and how that outcome is satisfied or defeated. The Nebraska community study supports the Lafayette County, Wisconsin, analysis, both areas having arrived at a high proportion of tenancy, in exposing the increase of tenancy as an outgrowth of the natural life processes of former operating land owners.

This Nebraska study, however, postulated further consequences: that is, the outcome of the institution of tenancy. In this phase of the work the study persistently suggests that the undesired consequences of tenancy, so often pointed out by other land tenure studies, either do not exist or give no cause for remedial action. However, in respect to these matters, there is either no attempt to offer evidential materials, or those which are presented are not conclusive on their own terms, or they fail to reveal in the relevant cases the sequence of experience that leads to the consequences claimed in the study.

LANDLORDISM

Just before the turn of the century, a predominant belief accompanying the apprehension created by Henry George and others was that the rise of farm tenancy was caused by a concentration of ownership in the hands of land monopolists. Reflecting this fear, the U.S. census in 1900 collected special data on farm landlords and reported that although a low percentage of landlords held a higher percentage of land acreage than their numbers would indicate, still there was no evidence of any marked concentration of ownership of rented farm units or of ownership by absentee, corporate, or alien monopolists.

[62] Robert Diller, *Farm Ownership, Tenancy, and Land Use in a Nebraska Community* (Chicago, 1941).

No further attention was given to the subject until after the 1919 land boom, which again raised speculation as to whether, in the rapid turnover of the period, the ownership of rented farms might have tended to concentrate. Consequently, the Division of Land Economics made special tabulations of supplementary 1920 census data on farm landlords. This quasi census study was published in two bulletins in 1926.[63] These bulletins describe various characteristics of farm landlords and show that northern landlords are mainly elderly retired farmers, whereas southern landlords are more frequently younger, active operating managers of groups of rented farms. Concentration of absentee, corporate, or alien ownership was found still to be inconsequential.

In 1924 Nebraska issued a bulletin devoted to landlordism because "The landlord has been ignored or given brief, passing, or incidental comment; and what has been written of landlords has not shown" any statistical basis.[64] In contrast to the Nebraska bulletins previously referred to, which constitute a long series of reports on one 1920 survey, a topic heading asks, "Why study landlords?" The answer is that with the growth of tenancy landlords represent a "great and growing power" which should be analyzed carefully. The study contains the same type of material in respect to landlords as the 1900 and 1920 studies of the federal agencies. It also includes descriptions of the contents of rental contracts.

When, in the depths of the depression of the early 1930's, farms by the thousands were foreclosed, various states issued bulletins showing the extent to which farm lands had become the property of corporations and discussing some of the problems and practices involved in corporate ownership.[65] In the West the bulletins reporting on land ownership have been concerned not only with the extent of corporate ownership but also with the extent of public ownership by federal agencies and by lesser units of government. Also, as was pointed out in Chapter VI, in the grazing areas the total problem of land utiliza-

[63] H. A. Turner, *The Ownership of Tenant Farms in the United States* (U.S.D.A. Bull. 1432, 1926); *The Ownership of Tenant Farms in the North Central States* (U.S.D.A. Bull. 1433, 1926). See also L. C. Gray *et al.*, *Yearbook, 1923*.

[64] J. O. Rankin, *Landlords of Nebraska Farms* (Nebraska Agr. Expt. Sta. Bull. 202, 1924).

[65] W. G. Murray and R. C. Bentley, *Corporate-Owned Land in Iowa* (Iowa Agr. Expt. Sta. Bull. 307, 1933); W. G. Murray and W. O. Brown, *Farm Land and Debt Situation in Iowa, 1935* (Iowa Agr. Expt. Sta. Bull. 328, 1935); W. G. Murray and H. W. Bitting, *Corporate-Owned Land in Iowa, 1937* (Iowa Agr. Expt. Sta. Bull. 362, 1937); G. H. Aull, *Rural Land Holdings in South Carolina* (South Carolina Agr. Expt. Sta. Bull. 331, 1940); A. A. Dowell, *Corporate-Owned Farm Land in Minnesota, 1936–1940* (Minnesota Agr. Expt. Sta. Bull. 357, 1942); Harold Howe, *Corporations as Landlords of Kansas Farms* (Kansas Agr. Expt. Sta. Agricultural Economics Report 17, 1943).

tion is so based on land tenure that it is nearly impossible to discuss them separately.[66]

LANDED DEBT

As the United States developed beyond the period of abundant unused land and past the generation that last had an opportunity to carve its own farms from the public domain, the necessity of borrowing funds with which to buy land increased. As was pointed out in Chapter II, hardly had the new land frontier disappeared before attention turned to the provision of public credit for farm purchasers. By 1916 Professor Thomas Nixon Carver of Harvard had prepared a popular bulletin telling farmers *How to Use Farm Credit;* national commissions had visited Europe and had made their reports; Carl Thompson had begun some farm credit studies at Minnesota and had published a survey of sources and costs of farm loans for the U.S. Department of Agriculture; and B. H. Hibbard at Wisconsin had issued the first experiment station bulletin on the subject.[67]

The Wisconsin bulletin is primarily a discussion of the need for credit, the availability of funds, the existing amount of debt, and interest rates charged for various types of loans. The discussion is supported by figures from courthouse mortgage records, state banking commission reports, and interviews with bankers and store merchants. The similarity in technique with that of Taylor's 1910 bulletin on renting (see above) and Ely and Hibbard's 1916 bulletin on credit for cutover land settlers (see Chapter V) is plain.

In 1916 the Federal Farm Loan Act established the federal land bank system. In 1921 the Division of Land Economics published a study of farm land purchase through land bank loans.[68] This study is based on the replies to a questionnaire mailed to borrowers from the federal land bank system. It represents an extremely effective use of this quick survey technique. The problem is whether the "credit permissible under [the farm loan act's] somewhat conservative provisions was made too small to effect a marked reduction in tenancy." Here is a specific problem posed in terms of the outcome of an experiment in action. The results showed that of the 8 per cent of farm loans held by the land banks, less than 15 per cent were borrowed for the pur-

[66] R. R. Renne, *Readjusting Montana's Agriculture, IV. Land Ownership and Tenure* (Montana Agr. Expt. Sta. Bull. 310, 1936), and *Montana Land Ownership* (Montana Agr. Expt. Sta. Bull. 322, 1936); Cruz Venstrom, *Railroad Grant Lands of Nevada* (U.S.D.A., 1940); Tom Rennard, *Landownership in Wyoming in 1935* (U.S.D.A., 1941).

[67] See J. D. Black, ed., *Research in Agricultural Credit* (Social Science Research Bull. 3, 1931); B. H. Hibbard and F. Robotka, *Farm Credit in Wisconsin* (Wisconsin Agr. Expt. Sta. Bull. 247, 1915).

[68] L. C. Gray and H. A. Turner, *Buying Farms with Land Bank Loans* (U.S.D.A. Bull. 968, 1921).

chase of land, and less than a third of these loans were to persons who were landless. These facts, together with an analysis of the second mortgages on land purchase transactions, form the basis for 7 pages of suggestions in a 27 page bulletin.

Clearly these suggestions are not tested by the evidence. They are not final scientific conclusions, but they carry a very high degree of suggestive force, in part because of the sharpness and specificity of the problem formulation and partly because all the data apply to a group of farmers who have proceeded along patterns of activity that by definition are relatively uniform for the problem at hand.

At the same time the department issued other bulletins on farm credit, but these analyses were made from the point of view of describing the character of the loans held by various lending institutions. Some of the early experiment station bulletins were also devoted to describing the features of various types of loans including loans other than those on real estate mortgages. As one bulletin put it: Since there is need for better understanding of loan sources, terms, and costs, "The main purpose of this study has been to collect some statistics and other information pertinent to these questions." The tendency to combine forms of credit other than landed debts into general credit studies was accentuated by the 1923 act establishing the federal intermediate credit banks.[69]

After 1923 credit became a large concern of agricultural economics, and land economics personnel devoted less attention to such matters — even to mortgage problems. Some of the early state credit studies did approach the problem, as did Gray in Bulletin 968, from the experience of the borrower. Although a crude North Dakota field study was issued in 1924, a more comprehensive and polished Minnesota field survey set the pace for several others in which attention was given to credit in relation to tenure and to the progress of farmers in acquiring equity.[70]

[69] C. O. Brannen, *Farm Mortgage and Commercial Bank Loans to Farmers in Arkansas* (Arkansas Agr. Expt. Sta. Bull. 208, 1926). For other early examples of the types mentioned, see V. N. Valgren and E. E. Englebert, *Farm Mortgage Loans by Banks, Insurance Companies and Other Agencies* (U.S.D.A. Bull. 1047, 1921); *Bank Loans to Farmers on Personal and Collateral Security* (U.S.D.A. Bull. 1048, 1922); and Leland Spencer, *An Economic Study of Rural Store Credit in New York* (Cornell Agr. Expt. Sta. Bull. 430, 1924); E. B. Brossard, *Rural Credits in Utah* (Utah Agr. Expt. Sta. Circular 48, 1923).

[70] R. E. Willard, *Report of a Farm Credit Survey* (North Dakota Agr. Expt. Sta. Bull. 175, 1924); B. M. Gile and J. D. Black, *The Agricultural Credit Situation in Minnesota* (Minnesota Agr. Expt. Sta. Tech. Bull. 55, 1928); B. M. Gile and A. N. Moore, *Farm Credit in a Plantation and an Upland Cotton District in Arkansas* (Arkansas Agr. Expt. Sta. Bull. 228, 1928); B. M. Gile, *The Farm Credit Situation in Southwestern Arkansas* (Arkansas Agr. Expt. Sta. Bull. 237, 1929).

Of these early field studies, two in Arkansas should be mentioned particularly because they illustrate so well one of the chief observations of this study. Although these two reports reproduce Federal Land Bank posters which proclaim that "the amortization plan of farm mortgage payment leads to the final elimination of the debt," and although in both studies some of the farmers had land bank mortgages and some did not, nevertheless data on "changes in financial status" of the surveyed farmers are given in a completely separate section toward the end of the report, and comparisons are made, not in terms of the type of credit used, but in terms of tenure classification or the decade in which land was purchased.

After 1929 farm credit studies were undertaken in increasing number. Aside from the studies that were specifically devoted to forms of credit other than real estate mortgages, the credit studies may be divided into four types.

Numerous mortgage studies were devoted to tracing the statistical trends in mortgage entries and releases from data obtained in county land records offices.[71] Others were made by analyzing the records of lending agencies.[72] A third type involved field record information. A few of these involved interviews with farmers specifically directed at their debt experience, but many of them make use only of data obtained in farm management account work and the like.[73] Studies dealing directly with the debt experience of farmers, rather than indirectly with the points in their experience available in public or institutional records, were not frequently made between 1930 and 1938.

Perhaps one reason why there is a hiatus in the previous three types of farm mortgage studies between 1932 and 1938 is that the

[71] E.g., V. R. Wertz, *The Farm Real Estate Mortgage Situation in Ohio* (Ohio Agr. Expt. Sta. Bull. 497, 1932); *The Farm Mortgage Situation in Putnam, Union, and Green Counties, Ohio* (Ohio Agr. Expt. Sta. Bull. 509, 1932); W. G. Murray, *An Economic Analysis of Farm Mortgages in Story County, Iowa 1854–1931* (Iowa Agr. Expt. Sta. Res. Bull. 156, 1933); E. H. Hinman and J. O. Rankin, *Farm Mortgage History of Eleven Nebraska Townships, 1870–1932* (Nebraska Agr. Expt. Sta. Res. Bull. 67, 1933); R. R. Renne, *Montana Farm Real Estate Mortgage Indebtedness* (Montana Agr. Expt. Sta. Bull. 383, 1940); Gabriel Lundy, *Farm Mortgage Experience in South Dakota, 1910–40* (South Dakota Agr. Expt. Sta. Bull. 370, 1943).

[72] E.g., F. F. Hill, *An Analysis of the Loaning Operations of the Federal Land Bank of Springfield from Its Organization in March 1917 to May 31, 1929* (Cornell Agr. Expt. Sta. Bull. 549, 1932); S. W. Warren, *Results of Farm-Mortgage Financing in Eleven Counties in New York State* (Cornell Agr. Expt. Sta. Bull. 726, 1939). See comments above in respect to the study by H. F. DeGraff.

[73] E.g., L. J. Norton, J. Ackerman, and C. R. Sayre, *Capacity to Pay and Farm Financing* (Illinois Agr. Expt. Sta. Bull. 449, 1938); E. E. Sparlin, *Farm Credit in Hempstead County* (Arkansas Agr. Expt. Sta. Bull. 399, 1940); *Farm Credit in Ashley County* (Arkansas Agr. Expt. Sta. Bull. 400, 1940); C. H. Merchant, *Farm Credit in Aroostook County, Maine* (Maine Agr. Expt. Sta. Bull. 418, 1943); Sargent Russell and A. H. Lindsey, *Agricultural Finance in Massachusetts* (Massachusetts Agr. Expt. Sta. Bull. 405, 1943).

growing pressure of farm debts, which created interest in credit problems during the 1920's, broke into a wave of outright foreclosures in 1933 to 1935. Consequently, in the 1930's, a few states issued bulletins dealing directly with farm foreclosures.[74] This widespread foreclosure movement undoubtedly explains the resurgence of land economists' interest in farm tenancy and land ownership during the 1930's. Also, it reestablished the interest of land economists in agricultural finance to which they had paid increasingly less attention since 1923. In view of the fact that since 1937 the federal government has provided a special farm credit program specifically designed to aid tenants to acquire ownership — a suggestion made by the Division of Land Economics in U.S.D.A. Bulletin 968 and in the *1923 Yearbook* — it is likely that land economists will show particular interest at least in this phase of current farm credit activities.

TAXATION OF LAND

Practically all land is held, even in a system of fee simple ownership, with the obligation to help support the government under whose jurisdiction the ownership is enjoyed. This feature of land tenure has always created a close association of interest between those who center attention on problems arising out of the right of men in land and those concerned with problems arising out of the support of government. As was pointed out in Chapter II, Professor Richard T. Ely always included land taxation as a phase of land economics important enough to warrant special attention. Among the rural social scientists, however, taxation did not attract research attention until 1925. At that time, the falling agricultural price level caused farmers to be aware of the pressure they felt from their fixed obligations, and especially those which were set in terms of the high price level of the World War I period. One of these, already covered above, was fixed charges on landed debts. Another was fixed charges in land taxes.

The pressure of farm real estate taxes in the 1920's was increasing because of the rapid advances in government services such as the construction of roads that accompanied the advent of the automobile. Also, in view of the degree of prosperity which prevailed in the cities and in view of the fact that some rural public facilities were enjoyed by urban residents, there was a feeling that the tax load was not only high, but was disproportionately borne by owners of farm land. In 1924 and 1925 Kansas published reports comparing farm and city

[74] E. C. Johnson, *Farm Mortgage Foreclosures in Minnesota* (Minnesota Agr. Expt. Sta. Bull. 293, 1932); W. G. Murray, *Corporate Land, Foreclosures, Mortgage Debt and Land Values in Iowa, 1939* (Iowa Agr. Expt. Sta. Res. Bull. 266, 1939); R. R. Renne, *Montana Farm Foreclosures* (Montana Agr. Expt. Sta. Bull. 368, 1939).

real estate tax levies and showing the trends in taxation over a long period of years.[75] The tenor of these reports and those which followed in other states is illustrated by the titles of two Wisconsin reports: *Tax Burdens Compared: Farm-City-Village* and *Who Pays for the Highways?*[76]

Just as the interest in farm credit broadened out in the latter half of the 1920's from landed debt to all forms of credit and credit agencies, so too, interest in taxation broadened from the burden of property taxes to all types of taxation and the whole structure of government.[77] Also, the new Division of Farm Finance in the Bureau of Agricultural Economics not only took chief responsibility for promoting research in farm credit but also in farm taxation. Few new areas of rural social science have expanded as rapidly in terms of the output of bulletins as did rural public finance after 1926.[78]

Rural public finance bulletins from 1926 on dealt not only with the relative tax burden of farmers and nonfarmers and with the historical trends in tax levies, but they also got into the subject of equitable tax assessments on farm properties. In fact, the number of reports which emphasized tax assessments on farm properties at least equals the number on taxation already referred to.[79] Characteristic of all the farm tax bulletins are the conclusions that a higher proportion of total tax revenues should be raised by forms of taxation other than the

[75] E.g., Eric Englund, *Assessment and Equalization of Farm and City Real Estate in Kansas* (Kansas Agr. Expt. Sta. Bull. 232, 1924); *Tax Revision in Kansas* (Kansas Agr. Expt. Sta. Bull. 234, 1924); *Trends of Real Estate Taxation in Kansas from 1910 to 1923* (Kansas Agr. Expt. Sta. Bull. 235, 1925).

[76] As examples, B. H. Hibbard and B. W. Allin, *Tax Burdens Compared: Farm-City-Village* (Wisconsin Agr. Expt. Sta. Bull. 393, 1927); H. W. Yount, *Farm Taxes and Assessments in Massachusetts* (Massachusetts Agr. Expt. Sta. Bull. 235, 1927); W. H. Dreesen, *Trends in Tax Levies in Oregon with Special Emphasis upon Rural and City Real Property* (Oregon Agr. Expt. Sta. Bull. 257, 1929); C. H. Hammar, *Missouri Farmer's Tax Position* (Missouri Agr. Expt. Sta. Bull. 291, 1930); F. P. Weaver, *The Rural Tax Problem in Pennsylvania* (Pennsylvania Agr. Expt. Sta. Bull. 263, 1931); Whitney Coombs, *Taxation of Farm Property* (U.S.D.A. Tech. Bull. 172, 1930); R. L. Thompson and B. W. Allin, *Louisiana Farm Taxes* (Louisiana Agr. Expt. Sta. Bull. 231, Pt. 2, 1933). Other comparable studies include at least 12 other reports between 1926 and 1934 from Colorado, Michigan, Missouri, Montana, New York, North Carolina, North Dakota, Ohio, South Carolina, South Dakota, Texas, and Virginia, and yet others in subsequent years from states including Alabama, Nebraska, Nevada, and Rhode Island.

[77] E.g., W. H. Yount and R. E. Sherburne, *The Cost of Government in Massachusetts* (Massachusetts Agr. Expt. Sta. Bull. 256, 1929); G. S. Klemmedson, *The Cost of Local Government in Laramie County, Colorado* (Colorado Agr. Expt. Sta. Bull. 361, 1930); K. H. Parsons, B. H. Hibbard, and A. J. Walrath, *State Aids and Rural Property Taxes in Wisconsin* (Wisconsin Agr. Expt. Sta. Res. Bull. 138, 1941).

[78] See J. D. Black, ed., *Research in Public Finance in Relation to Agriculture* (Social Science Research Council Bull. 1, 1930); M. M. Daugherty, "Review of Current Farm Taxation Research," *Journal of Farm Economics*, vol. XVI, 1 (Jan. 1934).

[79] E.g., Kansas Bulletin 232; M. M. Daugherty, *The Assessment and Equalization of Real Property in Delaware* (Delaware Agr. Expt. Sta. Bull. 159, 1928); W. H. Dreesen, *A Study in the Ratios of Assessed Values to Sales Values of Real Property in*

property tax, that government expenditures should be lessened or made more efficient, and that the methods of assessment should be made more equitable.[80]

Although all types of rural public finance bulletins have continued to appear ever since 1925, it is noticeable that the kinds previously mentioned loomed particularly large from 1927 to 1932 and since 1937. One explanation undoubtedly is that the farm tax problem shifted in 1932 and 1933 from one arising merely from inequalities in levies between farm and city and among farms to one that centered in the sharp rise in tax delinquencies and in the actual tax forfeiture of land at the depths of the depression. Consequently, the lack of surveys of tax inequalities in 1933–1936 does not reflect a decrease in emphasis on rural public finance problems but a shift to the more drastic form of these problems.

It has already been brought out in Chapters II and V that excessive tax delinquency and resulting strain on local government units struck in the pioneer areas of the northern Lakes States in the middle of the 1920's and was instrumental in speeding up the type of land utilization research that was in progress in that area. The work shifted from research to program planning, and the emphasis on governmental problems in land utilization studies was enlarged. Professor Hibbard's studies of tax delinquency were published in 1928 and 1929, and some of the general tax studies previously referred to have some references and data on tax delinquency, but the deluge came in the two years 1934 and 1935, when there appeared at least ten state bulletins specifically aimed at the subject of delinquent taxes. And from 1936 to 1944, at least eight additional tax delinquency reports were issued in seven states.[81]

Oregon (Oregon Agr. Expt. Sta. Bull. 233, 1928); R. W. Newton and W. O. Hedrick, *Farm Real Estate Assessment Practices in Michigan* (Michigan Agr. Expt. Sta. Special Bull. 172, 1928); G. B. Clarke and O. B. Jesness, *A Study of Taxation in Minnesota with Particular Reference to Assessments of Farm Lands* (Minnesota Agr. Expt. Sta. Bull. 277, 1931); C. H. Hammar, *The Accuracy and Flexibility of Rural Real Estate Assessment in Missouri* (Missouri Agr. Expt. Sta. Res. Bull. 169, 1932); E. H. Hinman, *Sales Value and Assessed Value of Nebraska Farm Land; 1921–1934* (Nebraska Agr. Expt. Sta. Res. Bull. 77, 1935); R. R. Renne and H. H. Lord, *Assessment of Montana Farm Lands* (Montana Agr. Expt. Sta. Bull. 348, 1937); Harold Howe and L. F. Miller, *Assessment and Collection of Farm Real Estate Taxes in Kansas* (Kansas Agr. Expt. Sta. Bull. 283, 1939); G. H. Aull, *The Sales Price and Assessed Value of Farm Real Estate in South Carolina* (South Carolina Agr. Expt. Sta. Bull. 334, 1941).

[80] The reports vary in their emphasis on these alternatives. For example, Louisiana Bulletin 231 emphasizes alternative revenue sources and specifically de-emphasizes reduction in expenditures; Massachusetts Bulletin 235 eliminates the possibility of revising the tax structure and puts emphasis on better assessment procedures. I. J. Call in *Farm-Property Taxation in New York* (Cornell Agr. Expt. Sta. Bull. 485, 1929), suggests all three.

[81] C. O. Brannen, *Tax Delinquent Rural Lands in Arkansas* (Arkansas Agr. Expt. Sta. Bull. 311, 1934); G. H. Aull and E. Riley, *Farm Real Estate Tax Delinquency in*

Just as the wave of mortgage foreclosures in 1932–1934 threw some emphasis in farm credit back upon land tenure, so too the wave of tax reversions put some emphasis in public finance back upon land tenure and utilization questions. For one thing, a good deal of the tax delinquency was on rural but nonagricultural land;[82] and it has always been the land economists among rural social scientists who have most often worked on the relations of farm and nonfarm aspects of the rural economy.[83] Another reason is, as has been shown in Chapters V and VI, that in the development of land utilization studies, increased attention was given to efficiency of community service as the end-in-view rather than to the level of family living.[84] An even more direct influence was the fact that in some areas, particularly in the grazing and cutover areas, the amount of land that reverted to public ownership was so great as to require public management decisions either as to its disposition to private title or its utilization as public property.[85]

South Carolina (South Carolina Agr. Expt. Sta. Bull. 298, 1934); W. T. Fulilove, *Tax Delinquency of Farm Real Estate in Fifty-Two Georgia Counties* (Georgia Expt. Sta. Press Bull. 406, 1935); R. C. Bentley and J. P. Himmel, *Tax Delinquent Farm Land in Iowa* (Iowa Agr. Expt. Sta. Bull. 325, 1935); W. O. Hedrick, *Farm Tax Delinquency in Michigan From 1928–1932* (Michigan Agr. Expt. Sta. Special Bull. 264, 1935); C. H. Hammar, *Land Tax Delinquency in Missouri* (Missouri Agr. Expt. Sta. Res. Bull. 224, 1935); W. P. Walker and A. B. Hamilton, *Tax Delinquency in Maryland* (Maryland Agr. Expt. Sta. Bull. 381, 1935); R. P. Callaway and P. W. Cockerill, *Tax Delinquency on Rural Real Estate in New Mexico* (New Mexico Agr. Expt. Sta. Bull. 234, 1935); C. E. Allred, P. B. Boyer, and R. E. Horne, *Tax Delinquency on Rural and Other Property in Tennessee* (Tennessee Agr. Expt. Sta. Circular 53, 1935); L. P. Gabbard, *Tax Delinquency on Farm Real Estate in Texas* (Texas Agr. Expt. Sta. Bull. 507, 1935); H. C. Grinnell, *Rural Real Estate Tax Delinquency in New Hampshire* (New Hampshire Agr. Expt. Sta. Bull. 290, 1936); R. R. Renne, *Readjusting Montana's Agriculture, VIII, Tax Delinquency and Mortgage Foreclosures* (Montana Agr. Expt. Sta. Bull. 319, 1936); Harold Howe, *Tax Delinquency on Farm Real Estate in Kansas, 1928 to 1933* (Kansas Agr. Expt. Sta. Circular 186, 1937); R. R. Renne and O. H. Brownlee, *Uncollected Property Taxes in Montana* (Montana Agr. Expt. Sta. Bull. 382, 1940); C. O. Brannen, *General Property Tax Delinquency* (Arkansas Agr. Expt. Sta. Bull. 311, 1934); W. H. Dreesen, *Rural Tax Delinquency Study of the State of Oregon* (Oregon Agr. Expt. Sta. Bull. 371, 1940); C. A. Bratton, *Tax Collection and Tax Delinquency; Fifteen Rural New York Counties* (Cornell Agr. Expt. Sta. Bull. 806, 1944).

[82] See R. B. Craig and O. J. Hall, *Tax Delinquency of Forest Land in Arkansas, 1932–1933* (Arkansas Agr. Expt. Sta. Bull. 340, 1937). Also note that just when agricultural economists were emphasizing tax burdens and unequal assessments of farm land, the Forest Tax Inquiry, mentioned in Chapter II, was also conducting extensive surveys on the same subject in respect to forest lands; see F. R. Fairchild and associates, *Forest Taxation in the United States* (U.S.D.A. Misc. Pub. 218, 1935).

[83] E.g., David Rozman, *Interrelationship of Land Uses in Rural Massachusetts* (Massachusetts Agr. Expt. Sta. Bull. 387, 1941); also Chaps. V and VI.

[84] Note particularly reference in footnote 32 in Chap. VI. Further indications of the extent to which the Division of Land Economics delved into public finance questions are: J. L. Spaulding, *Farm Taxes and the Cost of Public Services in Relation to Land Resources in Ringgold County, Iowa* (Iowa Agr. Expt. Sta. Res. Bull. 288, 1941); and O. O. McCracken and Frederick Arpke, *The General Property Tax Structure and Irrigated Agriculture in Pinal County, Arizona* (U.S.D.A., 1941).

[85] See O. J. Hall, *State-Owned Land in Arkansas* (Arkansas Agr. Expt. Sta. Bull. 370, 1939); R. J. Penn and C. W. Loomer, *County Land Management in Northwestern South*

Thus far in this discussion no mention has been made of methodological questions in connection with these many surveys of land taxation. The reason is that hardly any large series of bulletins is so nearly uniform in respect to techniques as the land taxation studies. Coming into existence as they did in 1925 and thereafter, tax studies coincided with the development of mass statistical techniques such as were emphasized in the 1928 handbook of the Social Science Research Council, described in Chapter III.

Furthermore, the data for taxation studies were of a type that could be easily collected in huge quantities. Every tax assessor's and county recorder's office constituted a veritable mine of quantitative data. With the public interest that taxation problems were attracting, with a peak of popularity in the use of mass statistical data, and with an infinite supply of such data at hand, it is not surprising that this avalanche of bulletins resulted. In addition, in 1933 and 1934, the devotion to this type of work was facilitated, as were the part-time farming studies, by work relief funds which made all the easier the assembly of these available figures.

Early in the work, attention was given to the construction of farm tax index numbers.[86] This led to greater interest in making time series analyses in which trends of assessments, tax levies, and collections could be correlated with changes in farm income, the price level, and so on. Similarly, data for any given year could be worked over by counties, by size and type of tracts assessesd, by type of levy, by assessment ratios, and on numerous other bases. In no other group of bulletins so far reviewed does one find more correlation coefficients, standard deviations, and coefficients of variation. In fact, only in rare instances among the other groups of bulletins reviewed can one find even arithmetic means with so much as a standard deviation.

What these studies all did was to meet the purpose, as stated in Kansas Bulletin 283, of testing inequalities in assessment among different areas, individual properties, and low and high value properties, and of presenting an analysis of the amount and character of tax delinquency. Or, in line with the statement in Iowa Bulletin 325, they succeeded in undertaking "to make available statistical data which summarizes the problem of unpaid taxes and to set forth and comment briefly upon some of the apparent economic relations that bear on

Dakota (South Dakota Agr. Expt. Sta. Bull. 326, 1938); R. B. Westbrook, *Tax Delinquency and County Ownership of Land in South Dakota* (South Dakota Agr. Expt. Sta. Bull. 322, 1938); M. H. Taylor and R. J. Penn, *Management of Public Land in North Dakota* (North Dakota Agr. Expt. Sta. Bull. 312, 1942).

[86] M. S. Kendrick, *An Index Number of Farm Taxes in New York and Its Relation to Various Other Economic Factors* (Cornell Agr. Expt. Sta. Bull. 457, 1926).

the amount, character, and geographical variations in tax delinquency."

If these studies did not result in specific reforms, it was not because of lack of variables. In Maryland Bulletin 381 the conclusion is that "there are so many factors associated with the extent of tax delinqency that it is difficult to ascribe the lack of paying taxes on time to any outstanding cause." Therefore, the suggestion is made that "taxpayers and county officials should study the conditions making for tax delinquency in their counties and endeavor to remedy such conditions."

It is hard to see how such conclusions could be satisfactory to the "taxpayers and county officials." After all, their doubts and perplexities were the problematic situation to begin with. Here is a good example of the emptiness of some social inquiry. But the work could still be satisfying to researchers who see their task merely as that of presenting orderly facts. Before the Farm Economic Association, for instance, Professor Daugherty concluded a review of farm taxation research with the observation that "taxation research has largely taken the form of *the gathering and presentation of data*. Now it may be said that this gathering and presenting is upon a scale *sufficiently large and sufficiently comprehensive to demand the respect* of all who come in contact with it." [87] These quotations illustrate clearly the existence of the Pearsonian concept of science in its most rudimentary form and the ineffectiveness of social inquiry conducted under its influence.

Land Values

The points that mark stages in land tenure processes are transactions in which certain rights in land are transferred from one person to another. It is therefore possible to construct an analysis of land tenure problems in terms of these transactions—the making and termination of rental contracts, sales, gifts, inheritance, the closing and release or foreclosure of mortgages, tax reversions, and the like. In all these ways, people allocate and change their relationships to each other through their rights in landed property. In this dissertation, reference to these forms of transfers of rights in land has been considered along with the problems growing out of the distribution of rights arranged in such transfers.

Generally, when economists speak of transactions by which things are exchanged among individuals, they think in terms of a market within which expendable commodities or services of a nearly homoge-

[87] M. M. Daugherty, "Review of Current Farm Taxation Research," *Journal of Farm Economics*, vol. XVI, 1 (Jan. 1934). Italics inserted.

neous and reproducible type are transferred for a money consideration. This concept has become one of the central ideas in economic thinking about activities in an economy that is in fact often characterized as a "market economy."

A good deal has been written elsewhere to indicate the difficulties in applying these notions of a market to property in land, which, physically as surface space and legally as real property, exists even beyond the life of the owner and is not consumable and far from homogeneous. Also, it is recognized that a high proportion of land transactions bear characteristics quite at variance with that of willing buyers and willing sellers meeting impersonally in business competition. Nevertheless, landed property, the last citadel of preindustrial society, has to a large extent been mobilized commercially in modern times.[88] Furthermore, because of the duration of time involved in commercial investments in landed property, special interest and importance have always attached to the determination of financial appraisals of property rights in land.[89] In a money economy such appraisals are made not only when a full ownership estate is transferred, but also when landed property is used as collateral for a money loan, as a basis for taxation, or as a source of rental income. Furthermore, some such appraisal undoubtedly is often made in connection with inheritance, the granting of gifts, or other transactions in which noncommercial motives loom large.

Although the U.S. Department of Agriculture issued two reports on land prices as early as 1906,[90] it was not until the land boom period of World War I that further special research dealt with problems arising from the valuation of rural real estate. This boom coincided with the establishment of the federal Division of Land Economics, and immediately studies were undertaken in Iowa and Kentucky, which were centers of this unusual land selling activity. The reports on these studies made by L. C. Gray, O. G. Lloyd, and G. W. Forster, all of whom had been students of Ely and Taylor at Wisconsin, were issued in 1920 and 1922.[91] The Iowa study was one of the first publications of the new Division of Land Economics.

[88] See Charles Abrams, *Revolution in Land* (New York, 1939), Pt. I; and Karl Polanyi, *The Great Transformation* (New York, 1944), Chap. 15.

[89] Cf. C. L. Stewart, *Some Economic Aspects of Farm Ownership,* (U.S.D.A. Bull. 1322, 1925), which reviews "records of cost and income" for "16 farms over 25 years" to show "buyers who must carry the cost and pay out the purchase price from income" that they "need to figure largely in terms of a full life span or generation."

[90] G. K. Holmes, *Changes in Farm Values, 1900–1905* (U.S.D.A., Bureau of Statistics Bull. 43, 1906); *Local Conditions as Affecting Farm Values, 1900–1905* (U.S.D.A., Bureau of Statistics Bull. 44, 1906).

[91] L. C. Gray and O. G. Lloyd, *Farm Land Values in Iowa* (Iowa Agr. Expt. Sta. Bull.

Both of these bulletins state clearly that they are aimed at discovering the causes or the economic and social forces behind the high prices and the unusual activity and finding the probable effect of the boom on the agricultural economy. In both cases, also, schedule data were obtained from people who participated in the sales which were investigated. But also certain "general information" was secured from "well-informed persons" and others "actually in touch with the situation," "data on net rents" were obtained in the field, and "farm management data" were collected in farm management surveys in the areas studied.

These studies obviously were not undertaken to prevent the boom. They were designed to clarify other persistent problems which the boom had confused—particularly the over-all land tenure issue of whether and how farmers could attain ownership of their land. Thus, the Iowa inquiry was interested in "the opportunity of farmers to acquire the land they cultivate" and the Kentucky report in "the economic and social status of the farming classes."

In view of the function of problem classification of these studies, it is important to note in both cases the probing nature of the manner in which various types of data were collected from different sources. Also significant is the fact that informal interviews with informed people directly in touch with developments are specifically described as a part of the procedure. Even in respect to the formally organized information, both reports include classifications of the motives or intents of buyers and sellers. Although these classifications are not extended into the rest of the inquiry, but remain apart as one item in a series of observations, they nevertheless indicate that the studies were trying to expose the behavior processes involved in the situation. It should be noted too that these reports contain facts from other sources, such as historical series from the census; and the Kentucky report even includes a calculation of farm ad lineage in a leading paper to show the extent of selling activity.

In both reports these different types of facts and information are strung together as evidence in an analysis that is conceptually well integrated. In both reports also, one notices that continual references are made to common beliefs, prevailing opinions, exaggerated notions, all of which indicates specifically the function of the reports to clarify social confusion.

As to conclusions, the two reports differ, probably because of the fact that two years lapsed between their publication. The Iowa study

874, 1920); G. W. Forster, *Land Prices and Land Speculation in the Bluegrass Region of Kentucky* (Kentucky Agr. Expt. Sta. Bull. 240, 1922).

was published before land prices had settled back very far. Its conclusions include a four-page summary explaining, insofar as possible, the actions taken by people in the land market during the boom, and a six-page discussion of the probable consequences ending with three specific questions for future analysis and a suggestion of a hypothesis in reply.

In the Kentucky report there was wariness of detailing the probable consequences as had been done in the Iowa report. On the other hand, it was possible to present "such effects as thus far have been registered" in financial losses. Consequently, a brief account is given of 146 cases in which lenders had to take action because of mortgage delinquencies. With analysis of the "income-producing power" of these particular farms unknown, "their ability to meet their financial obligations" is "conjectured" from a comparison with farm management data from a different survey with the concluding hypothesis that "If . . . the majority are to pay for their farms, the number of years in which payment may be made must be extended."

Interest in Kentucky in these cases did not end with the 1922 bulletin, for eight years later another report was issued dealing with the further experiences of 167 of the earlier buyers who had sustained losses in the interval.[92] This study is interesting as an example of following through on the results of previous work but it is inconclusive. Although the experiences of the boom buyers are followed out to some extent in subseries patterns (those who lost their places but still farm, those who left farming altogether, and those who still owned the places they had bought), yet there is no direction or purpose to the subgroupings, for no problem is posed to which the inquiry appears to be addressed.

Soon after the publication of the Iowa study new types of land value analyses appeared. These studies were not undertaken to examine how people act in respect to land transactions or to deal with such questions as how to expedite tenure progress. Rather, they were intended to offer aid to tax assessors, bank appraisers, and others who, in a period of rapidly changing land values, were at a loss to determine reasonable land valuations. The task was that of measuring the value of farm land.

The first study of this type was published at Missouri in 1921.[93] In it is devised "a plan for obtaining the agricultural value of Missouri

[92] Merton Oyler, *Social and Economic Effects of Land Speculation in Farm Families in Central Kentucky* (Kentucky Agr. Expt. Sta. Bull. 300, 1930).

[93] O. R. Johnson and R. M. Green, *The Agricultural and Market Value of Missouri Farm Lands* (Missouri Agr. Expt. Sta. Res. Bull. 179, 1921).

farm lands" on the basis of acres, yields, prices, and production costs. In a Minnesota bulletin of 1922, a "forecasting equation" is calculated "from which the probable sale price . . . of land . . . may be determined."[94] In this report income and productivity methods are rejected in favor of sales prices. The technique is to correlate actual sales prices reported per acre in an area with four characteristics of the lands sold: depreciated building cost, land class index, soil productivity index, and distance to market. Into the resulting regression equations, known factors on other farms are supposed to be inserted, and from their combination the value of the land in question is determined.

Although this technique does not produce very satisfactory results even when applied to the very data from which the equation was calculated, and although numerous major deficiencies are noted, the report suggests its general adoption with modifications for land valuation purposes.

This early use of formal mathematical correlation techniques was a forerunner not only of later uses of mass statistical procedures for land value and taxation data (see above), but also of its use in several other areas of rural social science, particularly in commodity price analysis work. It took place at Minnesota, where John D. Black had recently revitalized the Division of Agricultural Economics before its amalgamation with the farm management work of the Division of Agronomy and Farm Management, and Minnesota continued to be an institution where land prices work received comparative emphasis. These conditions and the fact that John D. Black was the editor of the 1928 handbook referred to in Chapter III are not mere coincidences.

In 1924 reports were issued in Texas and by the federal department in which statistical relationships between rents and land values were sought.[95] Federal Bulletin 1224 serves to clarify some of the questions raised in Chapter III above. The report is designed to obtain "a statistical measure" of "the theoretical relationship between land income and land value." While in the early part of this study emphasis is put simply on the "practical importance of a statistical measure of these relationships" to mortgagors and mortgagees, in the final section emphasis is placed on the bearing of the study on the

[94] G. C. Haas, *Sales Prices as a Basis for Farm Land Appraisal* (Minnesota Agr. Expt. Sta. Tech. Bull. 9, 1922).

[95] F. A. Beuchel, *The Relation Between Rents and Agricultural Land Values in Theory and in Practice* (Texas Agr. Expt. Sta. Bull. 318, 1924); C. R. Chambers, *Relation of Land Income to Land Value* (U.S.D.A. Bull. 1224, 1924).

problem of what to do to help farm tenants become farm owners under a credit program, and recommendations in this regard are suggested.

Bulletin 1224 is also of interest because it uses theoretical reasoning as offering subhypotheses for ordering data, and further, it recognizes the limitations of mass statistical relationships for testing hypotheses about human behavior. Note, for example, the following (italics inserted): "These *correlation coefficients* show, then, that average cash rents vary in response to variations in productivity as they probably would not do if custom were an effective force in their determination.

"*This is strong evidence in support of the theoretical conclusions* that competition and not custom is the effective force . . . *but it does not constitute a complete* statistical *proof* of it. These *high correlations* . . . *merely show* that when productivity is relatively high, cash rent is relatively high, and vice versa. But . . . the coefficients alone do not prove that cash rents are equivalent to farm rents. *If the process is kept in mind* by which average cash rents are maintained at approximate equality relative to productivity, it will be seen that average cash rents *must* be approximately equal to farm rents. There are two sets of forces by which this . . . equality . . . is maintained: (1) If cash rents become low . . . tenants will tend to move into it . . . and bid up the rents . . . But information is not generally available to tenants . . . Furthermore, moving is expensive and troublesome, so that *it is not probable* that the movement of tenants . . . keeps rents [relatively] equal . . . (2) Landlords will sell their farms rather than rent them if cash rents become lower than farm rents.

"Since it is not a very difficult matter for existing landlords or retiring farmers to sell their farms if contract rents are too low . . . it is *doubtless* [their] action . . . which has kept cash rents in approximate equality in relation to productivity. That is, their action *has been* the effective force in bringing about the high correlations. If this is true . . . then cash rents closely approximate farm rents."

This long quotation is given here because it clearly indicates that, in Bulletin 1224, although theoretical reasoning is used and modified in accordance with empirical facts and although the relations between the empirical facts at once affect the theoretical reasoning, beneath both of these types of generalizations there still is need to ground the analysis in terms of what people actually do. This report very well illustrates the joint interplay of concepts and empirical facts, yet it must be noted that in the quotation above (which is the key to the analysis and to the tenancy policy recommendations), nonetheless, the analysis lacks the final test which in this case requires information as to (1) whether tenants do hear about regional rent differentials and whether they do move upon receipt of such information and (2) whether retiring farmers and landlords do actually shift their investments back and forth between land and mortgages when farm rents and mortgage rates change position. Without this type of sequential data in respect to the actual processes of behavior, the analysis falls short of the most nearly final test that social science can ordinarily hope to apply.

By 1927 interest in land price statistics had developed sufficiently, along with the growing attention to the repercussions of continually falling land prices and unfavorable farm product prices, that the Division of Land Economics issued a circular on land value trends and undertook to establish an annual report on farm land transactions and the prices at which they took place.[96] With this work by Wiecking, the construction of land price indexes was shown to be feasible and some states undertook similar work.[97]

In other instances as early as 1928, historical trends in land prices were worked out and graphically related to other time series to show "various factors affecting farm land prices."[98] In addition, a few studies were made up to 1935 along the line of the early Missouri Bulletin 179 and Minnesota Technical Bulletin 9 "to determine quantitative relationships between selling price of land and the factors that

[96] E. H. Wiecking, *The Farm Real Estate Situation, 1926* (U.S.D.A. Department Circular 377, 1927), and *The Farm Real Estate Situation, 1926–27* (U.S.D.A. Circular 15, 1927). A series of circulars with similar titles is available through 1945, although in some cases several years' reports are combined into one.

[97] F. M. Thrum, *A Local Farm Real Estate Price Index* (Michigan Agr. Expt. Sta. Tech. Bull. 96, 1929); C. H. Hammar, *Missouri Farm Real Estate Situation for 1927–1930* (Missouri Agr. Expt. Sta. Res. Bull. 154, 1931); C. H. Hammar and R. P. Callaway, *Missouri Farm Real Estate Situation for 1930–1931* (Missouri Agr. Expt. Sta. Bull. 172, 1932); C. H. Hammar and R. K. Moore, *Missouri Farm Real Estate Situation for 1931–32* (Missouri Agr. Expt. Sta. Bull. 203, 1933); E. C. Johnson, *Farm Real Estate Values in Minnesota* (Minnesota Agr. Expt. Sta. Bull. 307, 1934); C. L. Stewart, *Farm Real-Estate Valuations in Illinois* (Illinois Agr. Expt. Sta. Bull. 399, 1934); A. A. Dowell, *The Trend in Sale Prices of Farm Real Estate in Minnesota* (Minnesota Agr. Expt. Sta. Bull. 338, 1938).

[98] W. C. Jenson and B. A. Russell, *Studies of Farm Land Prices and Ownership* (South Carolina Agr. Expt. Sta. Bull. 247, 1928); Harold Howe, *Farm Land Values in Kansas* (Kansas Agr. Expt. Sta. Circular 156, 1930); B. A. Russell, *Investigation of Farm Real Estate Values in Anderson County, South Carolina* (South Carolina Agr. Expt. Sta. Circular 50, 1933); E. H. Hinman, *History of Farm Land Prices in Eleven Nebraska Counties, 1873–1933* (Nebraska Agr. Expt. Sta. Res. Bull. 72, 1934); T. M. Adams, *Prices of Vermont Farm Real Estate* (Vermont Agr. Expt. Sta. Bull. 391, 1935); L. F. Garey, *Land Transfers in Twelve Counties in Nebraska, 1928–1933* (Nebraska Agr. Expt. Sta. Res. Bull. 107, 1938); J. H. Marshall, *Method of Payment and Comparative Prices for Tennessee Farms* (Tennessee Agr. Expt. Sta. Rural Res. Ser. 181, 1945); B. H. Luebka and J. H. Marshall, *Farm Real Estate Market in Tennessee, 1850–1944* (Tennessee Agr. Expt. Sta. Rural Res. Ser. 184, 1945).

affect" it, so that "quantitative measurements of land qualities may, in part, take the place of rough estimates of the degree to which different land qualities affect value." [99] About 1935, as was indicated in Chapter VI, these attempts to find statistical formulas for land appraisal work merged into the growing interest in land classification work, so interest shifted to drawing land maps instead of using mathematical devices to get accurate and standardized bases for land valuation, appraisal, and assessment purposes.[100]

One later analysis of the level of land values in a specific area should be noted because of its similarity to what has been said about U.S.D.A. Bulletin 1224. This report from Idaho is based on assessment data, mortgage record information, and tax assessments for several hundred land transfers.[101] The data are handled statistically to show a much narrower spread in land values than occurs in gross or net productivity. Out of this analysis some suggestions for action are arrived at; but again what is missing is information of the type necessary to test whether people actually do behave in accord with the highly suggestive conceptual explanation that is advanced.

Finally, with the advent of World War II, interest in land sales prices once more quickened as land prices rose; but in this period the federal Division of Land Economics at once launched a nationwide system to obtain quarterly statistical reports of activity in the farm land market.[102] It is doubtful, however, whether these materials, extensive as they are, will give as much insight into the processes that work to build a land value phenomenon as did the original Iowa study of 1920.

In this review of land value work is presented a very clear history of the influence of the research methodology concepts that carried weight around 1930 in many sectors of rural social science. Furthermore, the difference between the type of analytical work represented

[99] David Weeks, *Factors Affecting Selling Price of Land in the Eleventh Federal Farm Loan District* (California Agr. Expt. Sta. Hilgardia 3, 17, 1929); C. H. Hammar, *Factors Affecting Farm Land Values in Missouri* (Missouri Agr. Expt. Sta. Bull. 229, 1935); W. G. Murray and H. R. Meldrum, *A Production Method of Valuing Land* (Iowa Agr. Expt. Sta. Bull. 326, 1935).

[100] See *The Classification of Land* (Missouri Agr. Expt. Sta. Bull. 421, 1940), and Chap. VI above.

[101] A. N. Nybroten, *Land Values, Mortgages, Rents and Wheat Yields of Northern Idaho Wheat Lands* (Idaho Agr. Expt. Sta. Bull. 248, 1942). Also, *The Rural Land Market in the Northern Idaho Grain-Pea Area* (Idaho Agr. Expt. Sta. Bull. 261, 1945).

[102] E.g., N. J. Anderson, *What Price for This Land?* (South Dakota Agr. Expt. Sta. Bull. 368, 1943); D. E. Young, M. A. Brooker, and F. J. Welch, *Rural Land Market Activity in Mississippi* (Mississippi Agr. Expt. Sta. Bull. 406, 1944); Frank Miller and H. C. Filley, *Land Prices* (Nebraska Agr. Expt. Sta. Bull. 379, 1945); H. V. Stonecipher, Howard Mason, and Dora Dunn, *Wartime Land Market Activity in Northern Nevada* (Nevada Agr. Expt. Sta. Bull. 174, 1945). Also numerous mimeographed reports from state experiment stations and regional offices of the Division of Land Economics.

in the Iowa land boom study and that which was produced almost exclusively after 1922 illustrates the difference between (1) research which, though incomplete, sets out to clarify a problem and to point to suggested lines of action by seeking information as to what the people involved are actually doing, and (2) that which deals exclusively with the quantitative arrangement of a mass of residual facts which at best are only an index of what might have occurred.[103]

These reports also illustrate Deming's differentiation between type A and type B research problems. In both cases "the ultimate purpose is action." But in the statistical studies, which are type A, "*Interest centers in the product as it is,* not on how it got that way, or what it ought to be or might have been." The "problem" is that "of measuring something."[104] Put another way, it is a "problem" of "counting"—of counting those elements that have "previously been determined as strategic" to a previously determined type of action.[105]

But the land boom studies of the early 1920's are type B, in which "*Interest centers in the process,* the underlying cause system of forces . . . that give rise to yesterday's, today's, and tomorrow's product." In this case, "it is a matter of judgment and knowledge of the subject-matter to state the range of validity of a relationship, and to decide when enough situations have been covered to establish this validity with a sufficiently high degree of belief."[106]

These comments draw attention to another use these bulletins serve—as illustrations of an important point in methodological discussions recently clarified by Professor Kaufmann.[107] Kaufmann points out that the words *probable* or *probability,* key words in many discussions of scientific method, have three different meanings which it is important to distinguish. In one sense, *probable* merely has reference to the fact that in science a proposition is "never exempt from invalidation." In a closely related meaning, *probable* refers to the "degree of confirmation" of a proposition or to "probability preferences" in

[103] In the Social Science Research Council scope and method bulletin issued in 1933, the type of work represented by Gray and Lloyd's Iowa study is almost completely ignored. The field is described as consisting of "the theory and explanation of movements and compositions of values" (with emphasis on "theory") and of "methods and procedures for placing values on given pieces of farm realty" (with emphasis on "practice") and as being "most closely related to research in commodity prices." J. D. Black, ed., *Research in Farm Real Estate Values* (Social Science Research Council Bull. 19, 1933).

[104] W. E. Deming, "On a Classification of the Problems of Statistical Inference," *Journal of the American Statistical Association,* vol. 37, 177 (June 1942).

[105] L. A. Salter, Jr., "A Comment on Deming's Classification of Problems of Inference," *Journal of the American Statistical Association,* vol. 37, 220 (Dec. 1942).

[106] Deming, "On a Classification of the Problems of Statistical Inference," *Journal of the American Statistical Association,* vol. 37, 181 (June 1942).

[107] Felix Kaufmann, *Methodology of the Social Sciences* (New York, 1944), especially Chap. VI, "Truth and Probability."

accepting undecided propositions. It is with these meanings that Gray and Lloyd refer to the purpose of their study as determining "the probable effects" or "probable consequences" of the land boom. That is, by using the word *probable* they mean, first, that their statements may be invalidated by experience, and second, that the weight of their evidence confirms and leads them to prefer the conclusions they present.

In respect, however, to the statistical studies of masses of data, a third concept of probability is involved. In this case, from applied probability mathematics, statistical laws (which constitute "a specific type of rules of procedure relating to predictions of relative frequencies in large series of events") are followed to determine quantitatively the chances that the calculations made from the data at hand would be the same if additional calculations were made from other samples of the same data. While these metrical relationships may be helpful in deciding probability preferences for undecided propositions, they are not a necessary part of such decisions.

These distinctions by Kaufmann and Deming make clear some of the fundamental differences that may exist with social science materials, even though they are hidden under the blanket term *research* and may even be described and presented in apparently identical words. These distinctions also reemphasize the point in Chapter III that some of the Pearsonian formalities, introduced into rural social science after it got off to its own informal exploratory start, may have brought about, almost unnoticed, a misalignment between the original purposes of inquiry and the practices which were followed in the expectation of achieving those purposes.

Recapitulation

HISTORICAL

Research in problems arising from the distribution and holding of rights in land dates back to the earliest years of rural social science in this country. These problems are of such a type that they early attracted, and have consistently held, the attention of those who have stood in the political economy tradition which Ely fostered and in which he was followed by the figures of Henry C. Taylor and L. C. Gray.

The early publications clearly reveal the differences between the farm management group and the economists. The work at Wisconsin again shows disregard for quantitative precision and the placing of emphasis upon informal, but firsthand, information and conceptual reasoning. The work of others shows the direct influence of Warren's technique for farm management surveys and his analysis of survey data by gross quantitative comparisons between groups of farms.

It is noticed, however, that whereas public concern revolved around the apparent closing of opportunities for farm people to become full owners of the land they worked, Taylor, probably owing to the impression that his research in England made on him, stressed from the start the problem of arranging landlord and tenant relations so that tenancy would be less a source of private and social problems.

It is in the work of L. C. Gray particularly that some research interest remained trained on the question of instituting action to promote farm ownership by farm operators. It is also apparent that in Gray's research the informal Wisconsin approach is amalgamated with the growing respect for quantitative data precisely handled. The development of an almost exclusive regard for the compilation and arrangement of mass statistics is amply demonstrated by the huge amount of work which was undertaken on increasing scale in the 1920's and early 1930's in respect to land value and taxation data.

Also to be noted is the influence, largely arising from John D. Black's group at Minnesota, of combining the formal reasoning of neoclassical economic theory with the formal use of advanced statistical techniques in the analysis particularly of land price data.

The development of landed property research shows the close relation between investigational work and the pressure of existent social problems. Throughout the period there was some continued attention to making available to owners and operators of farms some information relative to their personal problems of drawing leases. But in respect to social problems, interest shifted from ownership and tenancy to land values and ownership transfers, to taxes and debts, to foreclosures and tax delinquencies, again to ownership and tenancy, and back to land values—as the country became successively aware of the existence of tenancy, a land boom, falling prices, severe agricultural distress, and signs of another wartime land boom.

METHODOLOGICAL

This review of landed property research bears out the discussion of scientific inquiry given in Chapter III. As compared with research in land utilization it shows a comparatively large amount of work, but much of it is channeled into one of a few set patterns of analysis, and there is relatively little progress so far as action recommendations from inquiry are concerned.

In the first place, a good deal of the work is not of the problem-solving type. Much effort has been given to describing existing lease forms and republicizing census data, not with any purpose of revealing sources of difficulty or finding solutions, but merely to make simple information available to any who might be interested in it. Only in Taylor's earliest work and in a few rare instances since, is there any evidence that investigations were specifically conducted for the purpose of clarifying difficulties and uncovering experiments in which these difficulties had been overcome.

On the contrary, with the increased formalization of techniques for collecting and presenting masses of data, there has been an increasing predominance of reports with no action problem posed, no problem explored, and no problem solved. Rather, there has been a large number of studies in which standard categories of farms have been described in respect to their averages and distribution for various items.[108] Instead of asking what action can be taken that will be consistent with this purpose, these studies ask: What is the quantitative relationship between this set of figures and that?[109]

In some cases there has been an attempt to pose problems for further analysis, but more often even this is not accomplished. In some instances, action suggestions are offered; but seldom are they supported by evidence arranged to show that if the recommended practices are effected, the postulated results are experienced.

While the tenure and tenancy studies have relied on tabular comparisons of averages, the land value, debt, and tax studies have more often used formal statistical correlations; but they have not been more effective in arriving at suggested actions that are shown to resolve problems. These deficiencies arise not merely from mistakes in the use of these statistical techniques, although such errors are common, but from the lack of formulated problems and hypotheses

[108] Cf. M. M. Kelso, "A Critique of Land Tenure Research," *Journal of Land and Public Utility Economics*, vol. X, 4 (Nov. 1934); Joseph Ackerman, "Status and Appraisal of Research in Farm Tenancy," *Journal of Farm Economics*, vol. XXIII, 1 (Feb. 1941); and G. S. Wehrwein in *Research in Agricultural Land Tenure*.

[109] While this study emphasizes the importance of the formulation of hypotheses and their use in sifting evidence, it is also clear from this inquiry that little help is available in a recent article on "Hypotheses in Land Tenure Research," since a hypothesis is there defined as "a tentative statement of a likely relationship between phenomena which, if verified under specific conditions, may be set forth as a principle or law." Under this ambiguous definition, which misses the whole point of action control in experiment, 20 hypotheses are listed. O. D. Duncan, *Journal of Farm Economics*, vol. XXV, 4 (Nov. 1943). In another article the same writer offers 15 other land tenure hypotheses for the use of sociologists. "A Sociological Approach to Farm Tenancy Research," *Rural Sociology*, vol. V, 3 (Sept. 1940). Cf. J. G. Maddox, "Land Tenure Research in a National Land Policy," Rainer Schickele, "Tenure Problems and Research Needs in the Middle West," and C. A. Wiley, "Tenure Problems and Research Needs in the South," *Journal of Farm Economics*, vol. XIX, 1 (Feb. 1937).

and the lack of evidence that specified practices result in specified consequences.

Without information arranged so that it points to a problem and integrated so that it reveals the actual processes of human experience, the inquiries at best result in weak suggestions, in the repetition of recommendations commonly expressed but seldom proved, or actually in the denial that anything can or should be done despite the acknowledgment of existent conflict and difficulty.

A few studies stand out because they center upon a problem, use diverse forms of information to explore its source and direction, and pose formulated hypotheses for further examination; or because, with the problem well formulated, they search out the cases — experiments — which indicate by what process — within what actual sequence of experience — the postulated purposes have been achieved or defeated.

Finally, these land tenure studies provide a clear illustration of the differences noted in Chapter III between procedures that reveal quantitative relationships among items in a mass of data and those which reveal the sequence relations among items in individual cases. Where data are grouped, for example, by "ladders," one knows the sequence of tenure experience in each case; but where data merely describe the average number of years of experience in various tenure stages and the average age at each stage, there is only a suggestion as to what might have actually been the experience of the subjects studied. But this review also shows that even in instances where the sequential experience of the subjects is revealed, this factor may be set apart as merely one of a number of descriptions of the material available. As a result, the work fails to seize the opportunity to associate strategic actions in these processes with their consequences in human experience. This failure is in large part explained by the fact that the object of those inquiries has been to summarize and present data in an orderly manner rather than to find cases that explain and test sequence of experience between purposes, actions, and outcome.

CHAPTER VIII

Changes in Emphasis

In the review of research work in land utilization in Chapters V and VI, it was made clear that such work concerned situations in which an area was undergoing a transformation in the sense that space once devoted to enterprises with certain land requirements was being put to uses with different requirements. To clarify this point, it was necessary to suggest a difference between a major land use and a minor land use. Although this question has been touched upon elsewhere, it is one that needs a little elaboration.

A region of land in a society will be divided, allocated among people, and assigned to some designated use. These units of space will be owned and occupied more or less in accordance with the needs of the type of activity in which the land users are engaged. Within any of these space units enterprises will be managed, and that management will, from time to time, put certain of its space units to different specialized uses. Thus, a farmer may sow one of his fields to grain one year, leave it idle another year, or put it in intertilled crops. Likewise, land in a residential unit may be devoted to a lawn at one time and to a driveway at another time. These are minor land use changes.

But if a farm unit is to be transformed to residential units, a different order of change is ordinarily involved because residential units have different land requirements from farming units. In the process of transformation, a shift in the whole pattern of economic space units is involved. A chain reaction is set up, for the establishment of these new major land use units affects the neighboring units directly and indirectly. Not only is there a reorganization of the physical layout of the space units; there is also a change in their tenure. And in the changes in the holding of rights in the old space units and the new, there is a different evaluation of the space. These new valuations result from a changed perspective on the physical elements that characterize that space. These physical elements are two in number: location or geometric relations, and tangible resources or physical qualities.

Thus, for example, when, during the war emergency, war plants were established where only farming had existed, there was not only a change in the size of the space units into which the land was divided and a change in the tenure of that space, but also the location of the space and the quality of the soil were evaluated in an almost wholly different light. Under such circumstances, whether the soil is good or excellent for crops is no longer a dominant point of consideration; whether the nearest shopping center or school is a quarter of a mile or a mile away may become a very important factor. All these changes affect not only the adjacent space units but all the units in the environs.

This illustration may serve to emphasize that where major land uses are shifting, a certain chain reaction is set up involving the whole complex of land factors. This is so whether the change is from a less to a more intensive type of use, as from grazing to crop farming, or from a more to a less intensive use, as from farming to forestry or from farming to no physical use at all—abandonment. It is also to be noted that what constitutes a category of major economic use will vary from place to place and from time to time. Thus, while it is common to think of farming as a major use distinct from forestry, grazing, residence, yet there will also be circumstances when a change from, say, wheat farming to dairying, or from general farming to specialized fruit farming would involve all the problems of any other type of major land use change.

Although any problem situation in land utilization, therefore, necessarily involves changes in the holding and evaluation of property rights, it is clear, as was seen in Chapter VII, that on the contrary, landed property problems can arise independently of a major land use change. Society can be and has been concerned with who holds what rights in space units, how those rights are distributed among individuals, and what valuations are placed on those rights, without regard for changes in the economic activities to which those space units may be dedicated. At different times and places the degree of interdependence between land utilization and land tenure processes will vary. Perhaps the highest degree of interdependence is that noted in the subhumid areas of the West where ranching is a predominant use.

In a similar manner, any interest in land tenure processes involves attention to the rights in space, which in turn are desired because of the locational or tangible resource qualities of that space. Yet at any given time, interest may center on the changing character of these qualities themselves rather than on the more general question of landed property problems or the yet broader issue of land utilization problems.

Location

In the literature of the whole field of land economics, attention has at times been centered on doubts and confusion in respect to the changing nature of the locational qualities of land. As might be expected, however, these questions have for the most part been dealt with as conceptual problems, and their relation to experienced problems has been mainly in respect to the location of cities, manufacturing enterprises, and transportation centers. Within cities there has been a good deal of attention, of course, to the locational characteristics of space units. This work, however, lies outside the scope of this inquiry.

In rural land economics there has been almost no work in this country in which the problems investigated have centered on the location of space units. The research referred to in Chapter VI in the section, "Urban Influences on Farmers," deals with problems arising from the locational characteristics of rural land holdings. It has been treated in the earlier chapter only because of its close connection with the actual process of transforming space from agricultural to urban types of uses.

Another example of work dealing with the locational characteristics of farms is a study of the influence of roads on agricultural lands in New York.[1] In view of the interest that rural people have at times shown in road improvement, it is surprising that more attention has not been given to investigations along these lines. But it should be recognized that a good deal of the land utilization work referred to in Chapters V and VI and some of the land values work mentioned in Chapter VII have included some analysis of locational problems and factors.[2]

One aspect of locational problems is that which deals with questions of freight rates and trade barriers. Thus far in the development of agricultural economics, such questions usually have been treated as a part of the field of marketing and pricing research. The direct local problems and changes associated with transportation costs have not received the attention of land economists except in theoretical dissertations or in urban and industrial land economics research. It may also be noted that there is a future possibility, particularly with the growth of public regulation of fluid milk markets, that the interest

[1] J. L. Tenant, *The Relationship between Roads and Agriculture in New York* (Cornell Agr. Expt. Sta. Bull. 479, 1929).

[2] T. E. LaMont, "Planning Location of Hard Roads and Electric Lines," *Journal of Farm Economics*, vol. XVIII, 3 (Aug. 1936). Also see Chap. V, footnotes 16, 24, 28, 29; Chap. VI, footnotes 4, 15, 39, 40, 41, 50, 72, 90, and 91; Chap. VII, footnotes 94, 99, and 101.

of land economists in problems arising out of the locational characteristics of farm units may coincide with the interest of marketing economists in problems arising out of the allocation of city milk supplies.[3]

Conservation

An investigation of the second basic characteristic of the space units on which human enterprises are conducted reveals that although considerable attention has been given to conceptual analyses of problems arising from the changing character of tangible resources, relatively little research attention of an empirical nature has been devoted to it by rural land economists. Still, as might be expected, rural land economists have given relatively more attention to problems of resource conservation than to questions of location, while for urban land economists, of course, the opposite is true.

Also it must be recognized that in many instances rural land utilization changes are engendered by alterations in the tangible resource qualities of the land. In other words, the point is that in many instances these alterations are so drastic as to involve complete changes in land utilization patterns, as has been shown in Chapters V and VI.[4]

Yet there are problems arising out of changing qualities of tangible rural land resources which have attracted attention but which are not so severe as to involve major economic land use changes. In fact, in recent years one of the most important programs of social action in the field of agriculture has been that of the Soil Conservation Service, and much of its work has involved some measure of public action. Yet most of its work has been done without significant changes in major land use pattern or in property relations in land.

In Chapter VII it was noted that, at least since 1914, one of the reasons why farm tenancy has been considered a source of difficulty is the belief that farm tenancy leads to uneconomic deterioration of the soil. Consequently, in many studies of land tenure some data are offered to show whether tenant farmers, more than owner-operators, engage in soil-depleting practices. In other studies the same type of comparisons are made among groups of tenants under different kinds of leases.

Very little research attention centered directly on soil erosion prob-

[3] See V. L. Hurlburt, "Some Aspects of Administrative Pricing as Related to Land Economics Research," *Journal of Land and Public Utility Economics*, vol. XX, 2 (May 1944).

[4] E.g., W. N. Sparhawk and W. D. Brush, *The Economic Aspects of Forest Destruction in Northern Michigan* (U.S.D.A. Tech. Bull. 92, 1929).

lems until after 1932. In this connection it is interesting to note that although soil depletion and erosion were usually listed among the resource problems of the nation during the Theodore Roosevelt conservation era, actually these topics were almost ignored in practice—at least in comparison with the actions that were instituted for forests, minerals, and arid lands.

The physical scientists did some research work on soil erosion and of course a great deal on soil fertility before 1933, but it was not until after that year that a concerted public attack was made to check soil erosion. With this event, some economic research was initiated having as its central focus the problem of checking soil erosion. The first such study was published in Iowa in 1935.[5]

Iowa Bulletin 333 has a relatively long introductory statement of the problem. In this statement the problem is formulated in these terms: "In any region where the majority of the population depends directly or indirectly on agriculture, the perpetuation of soil productivity is of vital concern to the community as a whole. If conditions develop which threaten the perpetuation of the soil's producing power . . . they are threatening the future of the community at the same time."

The problem is then further refined, conceptually, in terms that suggest a hypothesis. "Many forces are compelling or inducing the individual farmer to exploit his soil. . . . High debt burdens and low agricultural prices exert a strong financial pressure upon the farmer to get as much out of his soil as he can, regardless of how its future productivity is affected." Also, there are "leases and landlord-tenant relationships which are detrimental to the conservation of soil fertility." The hypothesis, then, is that given the problem of community survival, if farmers are relieved of the pressure of debt and if land tenure institutions are altered, soil will be conserved.

The study was conducted by obtaining farm schedules in nine blocks of land in a watershed in which the Soil Erosion Service was operating. Among other things, each farm was rated according to the degree of erosion of its crop land.

In the first section of the analysis the farms are grouped by acreage, by topography, and by type of farming. Averages of the erosion index and percentage of land in soil-depleting crops are compared among these groupings separately considered. Here is an illustration of the confusion caused by separate descriptions of items for all farms. The size analysis shows that small farms have a higher average degree of

[5] Rainer Schickele, J. P. Himmel, and R. M. Hurd, *Economic Phases of Erosion Control in Southern Iowa and Northern Missouri* (Iowa Agr. Expt. Sta. Bull. 333, 1935).

erosion than large farms. The analysis of type of farming shows that hog farms have a higher average degree of erosion than dairy farms. And a cross-classification of type of farming by size of farm shows that on the average dairy farms are smaller than hog farms. Clearly, what is needed is not a series of cross-classified averages of all farms, but a set of subseries patterns to segregate, for example, farms that are small and are dairy farms and are eroded, from those which are small and are dairy farms and are not eroded.

The second analysis deals with land tenure. In this it is shown that operators with long periods of occupancy are on farms with a lower average erosion index and a lower percentage of land in corn than are those with shorter periods of occupancy. In these comparisons owners and tenants are lumped together. Therefore, although it is shown that the recently occupied farms have a higher percentage of tenants and a higher average erosion index, there is no way of knowing, for instance, how recent owners compare with recent tenants.

In the discussion of landlord-tenant relations the erosion index is not used, but comparisons are made which show that cash and crop share tenants have more land in corn than stock share tenants, and that corporate owned farms have more corn land, by a slight percentage, than do farms owned by individual landlords.

The third analysis, dealing with indebtedness, consists of two tables which show that when farms are grouped by amount of mortgage debt per acre, the more heavily indebted farms have a higher percentage of land in corn. And for farms on rolling land "the [average] erosion rating increases steadily with increasing size of mortgage per acre." This last quotation is important because it is typical of the most common type of quantitative analysis in use in rural social science since Warren's first agricultural survey made use of comparisons of cross-classification averages.

The implication of this tabular cross-classification description is that if debt is increased on a farm, erosion increases. But this is quite a different statement from one that says that all farms with high debt at this moment have an average erosion index that is higher than that for farms with lower indebtedness. The latter statement may be suggestive of the former, but the materials presented do not constitute as strong an evidential test as they would if differently arranged.

This point is more obviously a source of error in some cases than others, depending upon the nature of the problem and the degree of differences among and the dispersions around the averages that are compared. In the present instance the point is particularly important because soil erosion is a condition that develops over time; and a description of tenure, debt, or erosion at a moment of time cannot be conclusive evidence of the processes in land tenure and debt experience which have led to the present condition of the soil resources.[6]

In Iowa Bulletin 333 there is no reticence in posing recommendations for action in respect to changes in leasing practices, a program for enabling farmers to increase the size of their farms, and measures to relieve the pressure of debt. In view of the type of factual materials presented, these recommendations must be regarded as suggestive only of further inquiry into the problems raised.

It should also be noted that in this bulletin the original formulation of the problem, that the soil must be saved if the community is to survive, is not questioned in the study or supported by evidence. This fact indicates that the formulation of the problem itself ought to be probed. Without such analysis, the study under review is less than complete. But it should be noted that even though the problem formulation is not tested, nevertheless the statement of the problem is made clear so that one who reads the report at least knows within what setting the analysis is relevant.

It is interesting to note that in a later Iowa bulletin on the same subject the pattern of analysis employed is the same as that used in Bulletin 333.[7] In the later report the importance of time sequence in tenure relations is recognized, and information is presented in respect to changes in ownership, tenancy, and leasing systems on the farms studied over a period of years. But even then this analysis is given in a separate section, so there is no direct tie-up between these sequential tenure experiences and soil depletion and erosion conditions.

In spite of the procedural difficulties in these Iowa bulletins, they still must be noted as exceptional studies because they both seek to explain the reasons why soils on farms are not conserved. In this respect they are different from most bulletins dealing with soil conservation problems.

In contrast with the two preceding reports, most soil conservation bulletins deal with internal farm business analyses, except that the farms may be grouped and statistically compared in the following

[6] Cf. O. T. Osgood, "Some Observations on the Relation of Farm Land Tenure to Soil Erosion and Depletion," *Journal of Land and Public Utility Economics*, vol. XVII, 4 (Nov. 1941), pp. 413–14.

[7] Rainer Schickele and J. P. Himmel, *Socio-Economic Phases of Soil Conservation in the Tarkio Creek Area* (Iowa Agr. Expt. Sta. Res. Bull. 221, 1938). Also see: *Problems of Land Tenure in Relation to Land-Use Adjustment* (U.S.R.A. Land Use Planning Publication 9, 1936), by the same authors; and Rainer Schickele, *Economics of Agricultural Land Use Adjustments, I. Methodology in Soil Conservation and Agricultural Adjustment Research* (Iowa Agr. Expt. Sta. Res. Bull. 209, 1937).

ways: (1) the farm data averages for the farm before a soil conservation program is introduced are compared with similar data after the program has been in effect,[8] or (2) farm data averages for farms that accept the soil conservation program are compared with similar data for farms outside of the program.[9] In some studies these two types of comparisons are combined.[10] Several studies of soil fertility and conservation have also been made which are even more confined to the point of view of conservation as a problem only of the individual farmer.[11] In fact, soil conservation problems in many cases may actually only involve the adoption of relatively minor shifts in soil management practices. In such instances, only the attention of research specialists in farm management are required.

If there is need for any other approach to soil conservation problems, it is in those areas where the changing quality of the physical resources constitutes an immediate public problem, where there are institutional obstacles to conservational farming, or where action other than that of managerial decision is required to attain the desired control of the soil. It is to these problems that land economists would be expected to direct their attention; but so far they have not. Surprisingly little research has been accomplished even though soil erosion has been accepted as a problem of national importance, and even though a whole system of local governmental units has been created to promote conservation programs in local areas. For the most part, the work that comes nearest to problems of this order is that of the

[8] E.g., F. D. Cornell, Jr., *A Social and Economic Survey of the Spencer Soil Conservation Area* (West Virginia Agr. Expt. Sta. Bull. 269, 1936), which bulletin only lays a foundation set of figures for possible later comparisons; M. J. Peterson, *An Economic Study of Agriculture in the Little Beaverdam Creek Area, Anderson County, South Carolina* (South Carolina Agr. Expt. Sta. Bull. 332, 1940).

[9] E.g., D. H. Walter, *An Economic Study of Farming in the Crooked Creek Area, Indiana and Armstrong Counties, Pennsylvania* (Pennsylvania Agr. Expt. Sta. Bull. 369, 1938); E. C. Weitzell, *Economic Implications of Soil Conservation in Marshall County* (West Virginia Agr. Expt. Sta. Bull. 293, 1939), and *Farm Management for Soil Conservation in the Harrison Area* (West Virginia Agr. Expt. Sta. Bull. 301, 1941). A. C. Bunce combined information of the same type with an opinion poll of the farmers' evaluations of the program of the Soil Conservation Service in *The Farmer Looks at Soil Conservation in Southern Iowa* (Iowa Agr. Expt. Sta. Bull. 381, 1939). This procedure is a useful device in helping to shape problems and hypotheses for further inquiry.

[10] E. C. Weitzell, *Economics of Soil Conservation in West Virginia* (West Virginia Agr. Expt. Sta. Bull. 305, 1942).

[11] E.g., W. L. Gibson, Jr., *An Economic Study of Farming in Appomattox County, Virginia* (Virginia Agr. Expt. Sta. Bull. 311, 1937); B. H. Pubols, A. E. Orr, and C. P. Heisig, *Farming Systems and Practices and Their Relationship to Soil Conservation and Farm Income in the Wheat Region of Washington* (Washington Agr. Expt. Sta. Bull. 374, 1939); J. W. Coddington and D. E. Derr, *An Economic Study of Land Utilization in the Tobacco Area of Southern Maryland* (Maryland Agr. Expt. Sta. Bull. 424, 1939); F. L. Morison and J. I. Falconer, *The Relationship between Soil Maintenance and Profitable Farming* (Ohio Agr. Expt. Sta. Bull. 604, 1939); W. L. Gibson, Jr., *Farm Management Aspects of Soil Conservation on Flue-Cured Tobacco Farms in Virginia* (Virginia Agr. Expt. Sta. Bull. 327, 1940).

Division of Land Economics under the program of flood control surveys in 1938–1941. But that work was undertaken as administrative investigations and has not been made generally available in the literature of land economics research.[12]

Finally, it is important to note a recent appraisal of the research work sponsored for a few years by the Soil Conservation Service for the purpose of checking on the benefits of their program — the type of work referred to above in footnotes 7, 8, and 9. Neil W. Johnson, who made an extensive investigation of this research work, found that too much time and effort were being spent on the collection and summary of ordinary farm business survey or farm account record data.[13] But he particularly has emphasized the relative futility of putting "too great a dependence on average income differences." Instead, he has urged "Studies of representative case farms over a period of years" and has emphasized the desirability of "segregating farms in groups that 'speak the same language'" — that is, to "see some similarity between the physical, economic, and social situations on farms, some problems shared in common in attaining conservation goals." "Having isolated the situation needing study, these should be reexamined in an attempt to confine analyses to those on which the impact of a conservation program is clearly different." Also in his recommendations is the suggestion of an "experimental approach" in which the research worker, after actually helping to devise a conservation plan for a farm, "undertakes to 'live into' the farm situation over a period of time, recording both quantitatively and qualitatively what happens as the conservation plan is carried out."

These conclusions of Johnson's analysis refer to the improvement of research in the farm management aspects of soil conservation programs, yet they point to research method and procedural concepts that are in line with those which have come to the surface in the earlier sections of the present critical review of land economics research.

The present investigation fails to uncover any quantity of land economics research dealing with problems which center in the changing character of locational or resource attributes of space units, where

[12] Even such studies as that of A. N. Garin and G. W. Forster, *Effect of Soil Erosion on the Costs of Public Water Supply* (U.S.D.A. SCS-EC-1, 1940), are found only in limited supply; but see A. N. Garin and L. P. Gabbard, *Land Use in Relation to Sedimentation in Reservoirs, Trinity River Basin, Texas* (Texas Agr. Expt. Sta. Bull. 597, 1941).

[13] N. W. Johnson, "Needed Developments in the Evaluation of Soil Conservation Benefits," *Journal of Farm Economics*, vol. XXIV, 1 (Feb. 1942); *Sorting and Sampling Farms for Soil Conservation Research* (U.S.D.A., 1939); and *Analysis of the Present Program of Research in the Economics of Soil Conservation and Suggestions for Its Improvement* (U.S.D.A., 1940).

major land use or tenure adjustments are not centrally involved. The lack of research of this type emphasizes the dominance in land economics of those problems which do center in major land use changes and in landed property relations. But it also reflects the closeness of land economics to current public issues, for during the larger part of the growth of land economics, public attention was not directed to soil erosion. Furthermore, during the period since soil conservation has come into public consciousness, action has been effected through a specialized agency dominated by physical scientists and engineers. And, as was earlier pointed out in respect to irrigation and other western land problems, in all of these instances little room has been left open for independent social science research on the public aspects of these issues and programs.

CHAPTER IX

Conclusions

Until the turn of the present century, United States land policies were based on the assumption that nearly all lands were suited to private ownership and control and primarily to farming use. It was believed that if the public lands were put into private hands the nation would have an ample supply of raw materials and the farmers would own their farms. This view received a sharp jolt about 1890 when, just as the nation entered an era of rapid development, it was realized that unappropriated resources were no longer abundant and that an increasing proportion of farmers were tenants.

The debate on this revolution in land policy coincided with the efforts of Richard T. Ely and other leaders in political economy to urge economists to give greater attention to existing social issues. The field of land economics in the United States stems from these developments.

Soon after 1900 Henry C. Taylor and other economists became interested in the economic problems of agriculture; a number of agricultural scientists began to study the business management problems of farmers; and religious and other groups initiated movements to enrich country life. Out of these activities came agricultural economics and rural sociology as fields of social science research.

The establishment of a center of interest in rural land economics is attributal to the work of Ely and his agricultural economics students, Henry C. Taylor and Benjamin H. Hibbard, and even more perhaps to their student, Lewis C. Gray, who directed a Division of Land Economics in the United States Department of Agriculture from its inception in 1919 until 1939. This area of work has dealt primarily with those problems of public or political economy which arise out of changes in the major forms of utilization of rural land and the distribution of rights in land.

The history of rural land economics research shows a close connection with current public issues. Before World War I the existence of tenancy and the availability of land for settlement took precedence in research interest. After the land boom following the war, attention shifted from tenancy to land valuation and ownership and to the burden of land debts and taxes that the boom had fostered. Also, with the decline in the demand for agricultural products, studies of land abandonment and of agricultural decadence replaced work on land settlement. Because of the overwhelming importance of these problems in certain localities, the idea developed that land utilization research was a basis for community organization and planning.

With the advent of the depression and the New Deal era of public action, land economics research emphasized the use of public land purchase, reforestation, rural zoning, and subsidized relocation as means for resolving local difficulties associated with settlement on isolated or poor quality land, tax delinquency, land abandonment, conflicting uses, and other phases of land utilization adjustment. Such work came to be seen as a possible basis for aligning various types of public activities into an integrated attack on rural problems, and for a few years prior to World War II, land use planning held a central place of interest in land economics interest.

The depression also resulted in widespread foreclosures of mortgages, and this phenomenon recreated an interest in the problems of farm tenancy comparable to that which existed before World War I. The parallel also extends into the World War II period, when interest shifted from farm tenancy to land values, and immediately following World War II, to attention to land ownership changes.

Throughout the history of rural land economics, its methodology has been affected by corollary professional developments, just as its content has been affected by changes in public issues. To the precedents set by Taylor in research procedures, adoptions from other branches of investigation have been added. Taylor relied heavily on conceptual reasoning, but he interwove it with qualitative summaries of his discussions with persons who were involved in the problems he probed and with graphic and geographic representations of a few statistics from secondary sources.

Land economics research has also been markedly affected by George Warren's farm management research procedures, in which quantified answers to standardized questions were summarized and compared by cross-classified averages and frequency distributions. In addition, land economists have incorporated procedures taken from classical statistics, from the field mapping work of soils scientists and geographers, from the social surveys of rural sociologists, from the budgeting technique of farm management research, and from the master plans of city planners.

Most land economics research has been undertaken with Ely's point of view, which emphasizes the German political economy tradition and also includes the use of concepts from English neoclassical economics. In addition, land economics has been affected to some extent by the private business management outlook of Warren and others who specialized in farm management work.

The amalgamation and modification of these public and private viewpoints and of Taylor's and Warren's initial research procedures with those of later date and from other fields has been accomplished by L. C. Gray, W. J. Spillman, and C. F. Clayton in the U.S. Department of Agriculture, by B. H. Hibbard and C. J. Galpin at Wisconsin, and by John D. Black and O. B. Jesness at Minnesota. A list of persons whose influence has been important in other respects would include, in addition to some of the above, many others, but especially George S. Wehrwein for his long devotion to the field in teaching and writing. Research contributions in special branches reflect the work of such men as William Allen and A. B. Lewis of Cornell (land abandonment studies and land classification techniques), Ernest Wiecking (land value statistics), H. A. Hockley (legal aspects of tenancy), and E. O. Wooton and R. P. Teele (the land problems of the arid regions).

But if rural land economics research has made use of contributions from other centers of research interest, it has also come to feel the deficiencies and to reflect the uncertainties that persist elsewhere in rural social science. The existence of these doubts and confusions among rural land economists in regard to their research posed the problem for this study. Both personal experience and a review of current land economics literature attest to the fact that research workers are in doubt as to what to do to get research results and make a real contribution to the solution of land economics problems. The initial hypothesis is merely that if changes are made in land economics research procedures, then more productive results will be achieved.

An exploration of the literature of rural social science research reveals that in all branches of the field the same doubts and confusions exist. It also shows that when rural social science was in a period of great expansion, late in the 1920's, the danger of unproductive research efforts was foreseen because of past experience in that direction. At the same time, therefore, special efforts were made to give the growing profession the best available information on research method; but careful analysis of this material reveals serious weaknesses in it. It is noted that the use of mass quantitative data is emphasized to an extreme degree and is given paramount status, even though in the same documents the highest commendations are given to several other research methods. There is no integration of these diverse appraisals into an integrated research concept. It is further noted that interest centers almost entirely on the techniques of handling collected information and that scant attention is given to the problems of research that precede the arrangement of collected data, even though these determinations are said to be of utmost importance.

These confusions suggest that something more than the refinement of techniques for summarizing collected data may be needed to resolve research confusion and that an inadequate conception of scientific method may be blocking consideration of important issues.

The conceptual formulation of scientific method in rural social science is Karl Pearson's, whose *Grammar of Science* is directly used as an intellectual framework and whose own work centered in the biological science in which many agricultural economists had themselves been trained. An analysis of Pearson's work reveals that his whole concept of scientific method is also restricted to that of summarizing already collected quantitative data, with wholly inadequate attention to the purposes of research, the formulation of research problems, or the outcome and consequences of research. This analysis substantiates the idea that something more than the adoption of more precise techniques for handling quantitative data may be needed, and suggests that something more than the Pearsonian conception of scientific method may be necessary to overcome the confusions in land economics research.

A review of methodological literature in the general area of social science reveals that a number of current treatises, rooted in Pearson, also fail to fill the deficiencies or to resolve the conflicts noted. It also reveals that in various branches of social science, confusions and debates similar to those in agricultural economics and in land economics are found in abundance.

In these disputes a chief division is between those who insist on the scientific precision of mass statistical techniques and those who are groping for some means in research to preserve sequences of human behavior more nearly as they exist in experience. In several instances this idea is connected with that of designing social research to be useful in respect to the control of human experiences. Implicit in such considerations are questions as to the purposes of social research, the problems to which it is directed, and the usefulness of its conclusions.

These issues as to the relevance of research problems and research conclusions to social action, it is noted, have also been a source of criticism in rural land economics and in agricultural economics, even though, as has been shown, the subject matter of the research has been very closely attuned to changes in public concern in current social problems. Furthermore, these questions fall precisely in those areas that remain conspicuously vague or absent in the Pearsonian conception of scientific method and in the expositions of those whose work rests on his formulation.

It is further noted that among those who advocate the de-emphasis of mass statistics in social research techniques and the re-emphasis of action in social research purposes, there remains doubt as to whether such shifts imply complete or only partial abandonment of scientific method in social inquiry. But in the recent literature of statistics in areas far afield from social science is found the beginning of a movement which involves a new conception of the application of metrical procedures when the purpose for which they are used is to control a physical process.

Furthermore, when attention is turned to recent analyses of scientific method by philosophers, it is found that current reappraisals of physical or natural science throw a new light on the whole subject of scientific inquiry. In John Dewey's work, and particularly in his *Logic,* there is found a modern, functional theory of inquiry, in which misconceptions in older views of scientific method are revealed, and in which the purposes, procedures, and consequences of experimental inquiry are joined into a unified comprehensive formulation.

Although the discussions of Dewey and other philosophers largely center in physical and biological research, the implication is that social research can learn from the experiences of these other fields. In some cases, sources of difficulty that have hampered progress in social inquiry are suggested. Furthermore, in view of the fact that predominant social research procedures are found to be based upon incomplete and outmoded formulations of scientific method, a reappraisal of social research concepts in the light of more modern and more thorough views of science as experiment should be undertaken before it is concluded that *social science* is a contradiction in terms.

Dewey's theory of inquiry is sufficiently comprehensive and flexible to provide a place for most considerations that have received support in discussions of social science methodology. By bringing these together within Dewey's formulation, somewhat modified, an outline of social inquiry is developed that appears to offer a means of resolving the points of confusion and conflict in respect to social research method, and that is consistent within itself and with experimental science in general.

The substance of this outline of social inquiry is (1) that social science deals with situations in living experience in which there is confusion and conflict with respect to knowing what to do to get specified results (problematic situation); (2) that by the interaction of reasoning and direct and indirect observation of experience a tentative selection is made of those aspects of behavior which appear strategic to the process of acting and getting results (problem formulation); (3) that by further interaction of reasoning and observation of experience a tentative proposal is made that if certain lines of action are instituted, then specified ends-in-view are attained (hypothesis); (4) that this initial hypothesis directs a search for evidence to indicate how it should be modified to be consonant with experience (processing of evidence); (5) that in this processing certain quantitative associations may be found between data bearing on actions and on consequences of action—associations that may be more or less suggestive of modifications in the hypothesis (evidence of relationships); (6) that it is necessary to exhibit the actual sequence between actions taken and results obtained within the experience of individual human beings in order actually to test the hypothesis and to warrant it as a conclusion (evidence of relations in experiment); (7) that the final test in social science is the unity between purposes sought and consequences experienced when the recommended action is taken.

A review of the variously suggested and often hotly debated procedures of social science indicates that this outline provides a place for each. It indicates that both the strengths and weaknesses claimed for these procedures are resolved if they are seen, not as alternative methods for the conduct of inquiry, but as procedures for processing various types of data which have different functions to perform in the conduct of a full social science inquiry by the scientific method of experimentation.

From this reasoning and probing of materials on social research method, the hypothesis is advanced that if the concepts in the above outline are applied in research in rural land economics, then research efforts will be more effective. In order to test this revised hypothesis, a critical review of published research on land economics problems was undertaken. The review was restricted, mainly but not exclusively, to printed research bulletins issued by the agricultural experiment stations and the United States Department of Agriculture.

In order to provide as complete a test as possible from past research experience, an exhaustive search of the literature of this research has been attempted. In the review, approximately 500 separate reports of research are covered. Sufficiently careful scrutiny of each of these is made so that they are grouped together on the basis of the problematic situation to which they appear to be addressed. They are further arranged on the basis of the pattern of objectives, materials, and conclusions that each represents and are so arranged and referred to in this study.

For those groups of research bulletins in which the objectives, materials, and conclusions have a very high degree of uniformity, comments are made on the group as a whole, or one or two reports are singled out for specific reference. In other instances, bulletins whose import is tangential to land economics problems are merely referred to as a group. In a few cases where a large number of studies have been subject to critical review elsewhere, reference is made to the group and to the results of the previous analysis. Throughout the review, however, an eye is kept for any unusual statements of problems, clear-cut hypotheses, special arrangements of evidence, or notable conclusions.

Approximately 125 individual research studies are singled out for separate analysis and comment. For one of the outstanding areas of land economics research, that dealing with land utilization in the Lakes States cutovers, a separate analysis for every bulletin published between 1916 and the present is presented in chronological order. By all these means, assurance is given that wherever past experience in rural land economics research offers an opportunity to modify and test the working hypothesis, such opportunity has been investigated.

For these reports, an attempt is made to determine the problematic situation with which the study deals, to see if a formulation of the problem has been shaped to characterize the working hypothesis if one appears to exist, and to check the summary and conclusions against these statements. Then, a review of the evidence is made to determine whether the conclusions appear warranted and to note how the hypothesis is related to the analysis.

This critical sifting of evidence, covering a forty-year period, touching every part of the United States and the whole range of problems that arise from major changes in the use or holding of rural land, uncovers no substantially full inquiries that satisfy the whole outline of inquiry presented above. It does, however, reveal pieces of research which succeed in clarifying issues and thus pose problems well formulated for further inquiry. It reveals some which establish highly suggestive quantitative relationships that stand in need of further testing by reference to the actual patterns of experience. Others neatly lead through a conceptual clarification of a problem with evidence that is drawn from sequential experience but is so poorly arrayed that others cannot even indirectly participate in the observations made; or again they attack clear-cut problems with specific and relevant evidence but fall short either of posing problems for further analysis or of advancing recommendations for action.

To the extent that various research undertakings have accomplished these steps, specific, definable progress has been made. Yet this review of land economics research also leads to the observation that in the absence of an outline of the requisites of full inquiry, succeeding studies fail to build one upon the other. If it is granted that practical considerations made it difficult frequently to engage in full inquiries, it becomes all the more important that segmental studies be so arranged that their specific place can be seen in order that other studies may add further steps in a progressive line of inquiry. Unfortunately, however, it is more common for pieces of research to be repetitive rather than progressional. And even where highly suggestive clues have been uncovered or where problems have been carefully defined in partial studies, these useful starting points have often lain idle for want of further connected study.

This analysis makes clear that one of the greatest obstacles to effective research is the persistent failure to pose a problem or a hypothesis as defined above. The objective of much research is no more clearly defined than "to present materials that may be of interest to others." With such a compass the outcome of the work is merely the presentation of a mass of data, the relevance of which to the problematic situation may be great or small but is not indicated in either case. Such work consistently fails to result in warranted recommendations for action or even in suggestive ideas for further inquiry. As previously explained, this type of work may not be without justification; but it cannot be regarded as an effective contribution to progressive social inquiry. In some cases, however, these reports carry recommendations for action; but it is clear that such recommendations come from some source other than the reported research.

This critique also indicates that in many instances problem formulations have become stereotyped, particularly after public programs have been inaugurated with such problems in mind. Under these conditions research fails to explore contrary clues and may actually overlook, as exceptional, evidence that suggests that an entirely different problem exists. In land economics research the most important example of a habitual problem formulation is that of commercial farming versus forestry in areas submarginal for agriculture.

This dissertation also shows that in some instances, techniques of handling data assume standardized patterns which neither check the formulation of the problem nor test a hypothesis. This source of difficulty is found particularly in connection with land classification. Instead of being regarded as a means of assisting exploration and analysis of a particular problem, a complex mechanical procedure comes to be taken as an end in itself. As a consequence, materials are sometimes prepared that are wholly irrelevant to the areas to which they are applied, or uses are claimed for which the procedure is inappropriate.

Other common instances of rigid techniques are found in the arrangement of statistical materials. Often, data collected in several areas are uniformly arranged according to the area in which they were taken, even though there is no indication that these areas differ significantly in respect to the problem at hand, and even though clues may be found in the data presented that suggest the existence of critically different but relevant patterns of human experience in all of the areas.

The practice of analyzing factual materials by describing separately or comparing various items which describe the whole, constitutes one of the greatest limitations of current analytical procedures. Even when such arrangements of data, either on maps or in columnar tables, appear to make important revelations, such disclosures are at best suggestive of inherent sequence in experience. But these procedures not only fail to prove the relations suggested, they often actually conceal whatever sequential evidence there may be in the cases (experiments) from which the data are drawn.

Instances are also found in which the observed materials are actually handled in such a way that the experimental character of the observed experiences is retained. Sometimes these data are simply left so they may be identified as cases by the reader, but they are not so treated in the research. Sometimes the importance of the sequence developed in the analysis is denied by the pattern of actions recommended in the conclusions. In other studies, however, cases which have similar patterns of experience are arranged in accordance with the formulation of the problem and the need for testing and revising the hypothesis; and in these instances the evidence poses definite suggestions for succeeding inquiry and offers suggestions for action that are supported by the analysis.

In reference to the historical development of research in rural land economics, Taylor's procedure had the merit of directly probing and clarifying existent problems and of bringing out suggestions that were found in experience. But its great limitation is its informality. Because it does not show the range of experiences tested, it is impossible for others to appraise the weight of evidence or to compare it with that obtained in other studies. As a result, the warrantability of the conclusions cannot be judged from the evidence presented. Likewise, the single case studies sometimes used in the federal department had the advantages of Taylor's procedure but accentuated its limitations even more.

Warren's procedure overcame these difficulties but had serious limitations of its own, at least when used by others. It drew observations from experience with metrical precision, but it so handled these data that their experimental character, and thus their testing force, was lost. Although Warren used some case references to check his gross relationships with actual experiences, succeeding workers retained his quantification techniques as the substance of inquiry. Over time, therefore, attention was diverted from the exploration of experience problems and the search for experimental actions that overcome these problems, toward the routine collection and presentation of quantified materials.

The introduction of classical statistical techniques and map comparisons provided additional tools for finding suggestive relationships, providing gross checks on ideas, and arranging and presenting research data; but it removed the investigator still further from the actual experiences of the subjects of investigation, attracted attention to mechanical procedures instead of to existent social problems, and often made easier the freezing of problem concepts, hypotheses, and conclusions.

The synthetic budgeting technique made possible the exploration of hypotheses when tests in experience were not available, but it may have lessened the search for such experiments as might have been found. The early work of rural sociologists with citizen committees, which was later duplicated in land economics, undoubtedly served to bring research into closer contact with existent problems, but it also probably lessened attention to the requisites of progressive, exploratory research. The rural sociologists have shown that the application of public action programs offers opportunity for experimental inquiry, and they have also urged land economists to remember that social research should center its attention on the experience of people. But the sociologists too have been caught up in the research concepts which prevail so that their inquiries have often failed to observe patterns of human experience as experimental tests of proposed actions.

These observations support and give point to the hypothesis, formulated after extensive probing of methodological literature and during years of firsthand experience in research activities in the field of rural land economics. This study is offered as an elaboration of that hypothesis to those who are specifically interested in finding further refinements applicable to the research problems in land economics which they face. The present inquiry itself illustrates the use of the outline of social inquiry it presents. The problematic situation to which it is addressed can well be described as a state of confusion as to what to do to get effective research results. The probing of this situation has led to a formulation that involves a basic reorientation of social research concepts to make them consonant with the experimental character of modern scientific method. In view of this formulation, factual materials, conceptual analysis, and experience have suggested an outline of social inquiry that appears to offer a way of resolving the confusion. This hypothesis has been tested and modified and expanded by the analysis of several hundred research bulletins which together describe the purposes, present the analyses, and pose the conclusions of the whole range of forty years of experience in rural land economics research in the United States. It is therefore presented here as an aid in directing future rural land economics inquiry so that the outcome of such activities may be in accord with the purposes for which the research is instituted.

In the future, the most important needed step is the adoption of a more comprehensive conception of social science inquiry. Research must be viewed in terms of its relevance to action. To do so means that the purposes and consequences of inquiry are given greater attention and that the mechanics of research are regarded not only as procedures for gathering data but as ways and means of observing processes of human experience as operating experiments.

Secondly, it should be recognized that research has its roots in problematic situations; that is, it exists because of conditions under which there is doubt as to what people should do because there is conflict between the purposes they are striving to achieve and the consequences they are experiencing. There is need for sharper attention to the preliminary exploration and clear definition of problems — that is, to the statement of these doubts and conflicts. In rural land economics these problems will arise in connection with the establishment of new major forms of the utilization of space, with the development of landed property relations among men, or with the changing character of locational or resource qualities of the space which men control.

The next step is to encourage the functional use of hypotheses. Hypotheses are suggested alternative lines of action that will lead to the achievement of purposes. Their function is to direct the search for evidence as to what might be done. The aim of research is not just to affirm or deny a hypothesis, but to expand and modify it until it represents warranted assertions, grounded in experience, as to what actions will result in a satisfactory pattern of major land uses, will create desirable landed property relations, will overcome the problems engendered by changing attributes of space.

A fourth step in the improvement of research in rural land economics is recognition of the limitations and advantages of various forms of factual materials as evidence. Scientific investigators must stand ready to make use of all types of data and to judge their accuracy not merely in terms of metrical precision but on the basis of how well they reveal patterns of actual human experience. Secondary statistics, quantified data from interview schedules, qualitative information, newspaper files, public documents, participant observer reports, local histories, all kinds of maps — among these and other forms of data there should be no a priori choice, except that only information which is sequentially arranged can positively substantiate reported patterns of experience. The goal in respect to evidence is to examine the full range of experience in which alternative lines of action have been tried. In this view, exceptional cases become not merely extremes to be cancelled out, but potentially useful proving grounds and potential sources of new suggestions. The basic form for the presentation of research results will finally shift from successive discussions of various items, factors, or elements as they affect all the subjects of the study, to discussions of the various lines of actions and sequences of experience revealed in the evidence.

INDEX

Made in United States
Orlando, FL
27 October 2025

71481474R00148